Inverse Problems - Recent Advances and Applications

Edited by Ivan I. Kyrchei

Published in London, United Kingdom

Inverse Problems - Recent Advances and Applications
http://dx.doi.org/10.5772/intechopen.102219
Edited by Ivan I. Kyrchei

Contributors
Neil K. Chada, Gregor Moeller, Michael Conrad Koch, Kazunori Fujisawa, Akira Murakami, Ivan I. Kyrchei, Abdur Rehman Armath

First published in London, United Kingdom, 2023 by IntechOpen
IntechOpen is the global imprint of INTECHOPEN LIMITED, registered in England and Wales, registration number: 11086078, 5 Princes Gate Court, London, SW7 2QJ, United Kingdom

British Library Cataloguing-in-Publication Data
A catalogue record for this book is available from the British Library

Additional hard and PDF copies can be obtained from orders@intechopen.com

Inverse Problems - Recent Advances and Applications
Edited by Ivan I. Kyrchei
p. cm.
Print ISBN 978-1-80355-222-4
Online ISBN 978-1-80355-223-1
eBook (PDF) ISBN 978-1-80355-224-8

Meet the editor

Dr. Ivan I. Kyrchei received an MSc in mathematics from Ivan Franko National University, Lviv, Ukraine in 1992, a Ph.D. from Taras Shevchenko National University, Kyiv, Ukraine in 2008, and a DSc in algebra and theory numbers from the NASU Institute of Mathematics, Kyiv in 2021. He is currently the Lead Researcher at NASU's Pidstryhach Institute for Applied Problems of Mechanics and Mathematics (IAPMM), Ukraine, where he has worked for more than 20 years. He has published more than 100 papers in scientific journals and international conference proceedings on the theory of quaternion matrix equations, generalized inverse matrices, and functionals of matrices over quaternion algebras. He is also the editor of three books and a member of the editorial boards of five international scientific journals.

Contents

Preface

Inverse Problems - Recent Advances and Applications examines new aspects of the mathematical modeling of inverse problems, their applications in physical systems, and the computational methods used. The book's five chapters are divided into two sections. Section 1, "Modeling and Formulations of Inverse Problems", discusses new approaches in the modeling and formulation of inverse problems in some physical systems. Chapter 1 formulates the initial statement of the inverse problem. Chapter 2 introduces recent applications of the Ensemble Kalman Filter to inverse problems, known as ensemble Kalman inversion. The subject of Chapter 3 is the evaluation of gradients in inverse problems where spatial field parameters and geometry parameters are treated separately.

Section 2, "Some Computational Aspects", concerns mathematical methods of solving some inverse problems. Chapter 4 reviews individual solutions for the tomographic problem, including strategies for removing deficiencies of the ill-posed problem by using truncated singular value decomposition and the L-curve technique. Chapter 5 discusses the least norm of the solution to some system of quaternion matrix equations and its expression by determinantal representations (analogs of Cramer's rule) using row-column noncommutative determinants.

Ivan I. Kyrchei
Pidstryhach Institute for Applied Problems of Mechanics and Mathematics,
National Academy of Sciences of Ukraine,
Lviv, Ukraine

Section 1

Modeling and Formulations of Inverse Problems

Chapter 1

Introductory Chapter: Some Preliminary Aspects of Inverse Problem

Ivan I. Kyrchei

1. Introduction

Physical research in science can be divided into two groups. The first is that when by complete description of a physical system, we can predict the outcome of some measurements. This problem is called the modelization problem or the forward problem. The second group of research consists of using the actual result of some observations to infer the values of the parameters that characterize the system. It is the inverse problem, which starts with the causes and then calculates the effects. The importance of inverse problems is that they tell us about physical parameters that we cannot directly observe.

2. Primary equations of inverse problem

The inverse problem is that one wants to determine the model parameters p that produce the observed data or measurements d. $\mathbf{F}$ stays for some measurement operator that maps parameters in a functional space $\mathfrak{P}$, typically a Banach or Hilbert space, to the space of data $\mathfrak{D}$, typically another Banach or Hilbert space.

$$d = \mathbf{F}p \quad \text{for } p \in \mathfrak{P} \text{ and } d \in \mathfrak{D}. \tag{1}$$

Solving the inverse problem amounts to finding point(s) $p \in \mathfrak{P}$ from knowledge of the data $d \in \mathfrak{D}$ such that Eq. (1) (or its approximation) holds. In the case of a measurement, operator is linear and there is a finite number of parameters, Eq. (1) can be written as a linear system, where $\mathbf{F}$ is the matrix that characterizes the measurement operator, and $\mathfrak{P}$ and $d \in \mathfrak{D}$ are corresponding vector spaces. Such inverse problem is called linear.

Inverse problems may be difficult to solve for at least two different reasons:

1. Different values of the model parameters may be not consistent with the data;

2. Discovering the values of the model parameters may require the exploration of a huge parameter space.

If it is acquired enough data to uniquely reconstruct the parameters, then the measurement operator can be injective, which means

$$\mathbf{F}(p_1) = \mathbf{F}(p_2) \quad \Rightarrow \quad p_1 = p_2 \text{ for all } p_1, p_2 \in \mathfrak{P}. \tag{2}$$

When $\mathbf{F}$ is injective, one can construct an inversion operator $\mathbf{F}^{-1}$ mapping the range of $\mathbf{F}$ to a uniquely defined element $\mathfrak{P}$. In the case of a linear inverse problem, $\mathbf{F}^{-1}$ is an inverse matrix. Further, the main features of the inverse operator are characterized by stability estimates that quantify how errors in the available measurements translate into errors in the reconstructions. It can be expressed as follows:

$$\|p_1 - p_2\|_{\mathfrak{P}} \leq \alpha \|\mathbf{F}(p_1) - \mathbf{F}(p_2)\|_{\mathfrak{D}}. \tag{3}$$

Where $\alpha : \mathbb{R}_+ \to \mathbb{R}_+$ stay for an increasing function, such that $\alpha(0) = 0$. This function gives an estimate of the reconstruction error $\|p_1 - p_2\|_{\mathfrak{P}}$ based on the error in the data $\|\mathbf{F}(p_1) - \mathbf{F}(p_2)\|_{\mathfrak{D}}$. When the reconstructed parameters are acceptable, for instance when $\alpha(x) = Cx$ for some constant C, then the inverse problem is called well-posed. When the reconstruction is contaminated by too large a noisy component, then the inverse problem is ill-posed.

Injectivity of $\mathbf{F}$ means satisfying the two conditions for a well-posed problem suggested by Jacques Hadamard [1], *Existence* and *Uniqueness* of solutions. Eq. (3) is the third Hadamard's condition, which is *Stability* of the solution or solutions.

Typically, inverse problems are ill-posed. Even when we have a linear inverse problem with invertible matrix $\mathbf{F}$, it gives an ill-posed problem that can be solved by using the Moore-Penrose inverse matrix [2, 3] and least squares solutions inducted by it.

The goal of many experiments is to infer a property or attribute from data that is indirectly related to the unknown quantity. Parameter estimation problems usually satisfy the first criterion of well-posed problems, since something is responsible for the observed system response. Instead, they violate the third criterion and" almost" violate the second criterion because many different candidate solutions exist that, when substituted into the measurement model, produce very similar data. The condition of stability is often violated, because the inverse problem is represented by a mapping between metric spaces, but inverse problems are often formulated in infinite dimensional spaces. Therefore, limitations to a finite number of measurements, and the practical consideration of recovering only a finite number of unknown parameters may lead to the problems being recast in discrete form. In this case, the inverse problem is typically ill-conditioned and a regularization can be used. One of the most famous regularizations is the Tikhonov regularization [4]. The idea of Tikhonov regularization may be introduced as follows. In its simplest form, it consists in replacing the Eq. (1) with the second kind of equation

$$\mathbf{F}^* \mathbf{F} p + \alpha p = \mathbf{F}^* d \tag{4}$$

where α is a positive parameter. It leads to that the problem of solving Eq. (4) is well-posed.

Unlike parameter estimation, inverse problems often violate Hadamard's first criterion since an optimal design outcome may be specified that cannot possibly be produced by the system. On the other hand, the existence of multiple designs (solutions) that produce an acceptable outcome violates the second criterion. From these, it follows inverse problems that are mathematically ill-posed due to an information deficit. In the parameter estimation case, the measurements barely provide sufficient information to specify a unique solution, and in some cases, the data could be explained by an infinite set of candidate solutions. Information from measurement data and prior information can be combined through Bayes' equation to produce estimates for the Quantities-of-Interest (QoI). In this approach, the measurements, d, and the QoI, $x \in \mathfrak{P}$, are interpreted as random variables that obey probability density functions (PDFs). The PDFs are related by Bayes' equation

$$P(x|d) = \frac{P(d|x)}{P(d)} P_{pr}(x) \tag{5}$$

where $P(d|x)$ is the likelihood of the observed data occurring for a hypothetical parameter x, accounting for measurement noise and model error ("likelihood PDF"), $P_{pr}(x)$ defines what is known before the measurement takes place about a hypothetical parameter x, ("prior PDF"), $P(x|d)$ is the posterior PDF, which defines what is known about x from both the measurements and prior information, and $P(d)$ is the evidence, which scales the posterior so that it satisfies the law of total probabilities.

Therefore, "*the most general theory is obtained by using a probabilistic point of view, where the a priori information on the model parameters is represented by a probability distribution over the" model space". A priori probability distribution is transformed into the a posteriori probability distribution, by incorporating a physical theory (relating the model parameters to some observable parameters) and the actual result of the observations (with their uncertainties)*" [5].

Author details

Ivan I. Kyrchei
Pidstrygach Institute for Applied Problems of Mechanics and Mathematics, NASU, Lviv, Ukraine

*Address all correspondence to: ivankyrchei26@gmail.com

References

[1] Hadamard J. Lectures on Cauchy's Problems in Linear Partial Differential Equations. New Haven: Yale University Press; 1923 (Reprinted by Dover, New York, 1952)

[2] Moore EH. On the reciprocal of the general algebraic matrix. Bulletin of the American Mathematical Society. 1920; **26**(9):394-395. DOI: 10.1090/S0002-9904-1920-03322-7

[3] Penrose R. On best approximate solution of linear matrix equations. Proceedings of the Cambridge Philosophical Society. 1956;**52**(1):17-19. DOI: 10.1017/S0305004100030929

[4] Tikhonov AN, Arsenin VY. Solution of Ill-Posed Problems. Washington: Winston & Sons; 1977. ISBN 0-470-99124-0

[5] Tarantola A. Inverse Problem Theory and Methods for Model Parameter Estimation. Philadelphia: SIAM; 2005. ISBN 0-89871-572-5

Chapter 2

A Review of the EnKF for Parameter Estimation

Neil K. Chada

Abstract

The ensemble Kalman filter is a well-known and celebrated data assimilation algorithm. It is of particular relevance as it used for high-dimensional problems, by updating an ensemble of particles through a sample mean and covariance matrices. In this chapter we present a relatively recent topic which is the application of the EnKF to inverse problems, known as ensemble Kalman Inversion (EKI). EKI is used for parameter estimation, which can be viewed as a black-box optimizer for PDE-constrained inverse problems. We present in this chapter a review of the discussed methodology, while presenting emerging and new areas of research, where numerical experiments are provided on numerous interesting models arising in geosciences and numerical weather prediction.

Keywords: ensemble Kalman filter, Kalman filter, inverse problems, parameter estimation, data assimilation, optimization

1. Introduction

Inverse problems [1–3] are a class of mathematical problems which have gained significant attention of recent. Simply put, inverse problems are concerned with the recovery of some parameter of interest from noisy unstructured data. Mathematically we can express an inverse problem as the recovery of $u \in \mathcal{X}$ from noisy measurements of data $y \in \mathcal{Y}$, expressed as

$$y = \mathcal{G}(u) + \eta, \tag{1}$$

where $\mathcal{G} : \mathcal{X} \to \mathcal{Y}$ is the forward operator, and $\eta \sim \mathcal{N}(0, \Gamma)$ is some form of additive Gaussian noise. Specifically $\mathcal{N}(0, \Gamma)$ denotes a normal distribution with mean 0 and variance Γ. Commonly the covariance can be taken to be some form of the identity, i.e. $\Gamma = \gamma^2 I$, where $\gamma \in \mathbb{R}$ is some constant and I is the identity. Inverse problems are of high interest due to the amount of relevant problems that arise in wide variety of applications, most notably geophysical sciences, medical imaging and numerical weather prediction [4–6]. The classical approach to solving inverse problems, which is the theme of this chapter, is to construct a least-squares functional, and the solution is represented as a minimizer of some functional of the form

$$u^* := \arg\min_{u \in \mathcal{X}} \frac{1}{2} \|y - \mathcal{G}(u)\|_\Gamma^2 + \lambda R(u), \tag{2}$$

where $\lambda > 0$ is a regularization parameter and $R(u)$ is some regularization term, usually required to prevent the overfitting of the data. A common example is Tikhonov regularization, i.e. $R(u) = \frac{1}{2}\|u\|^2$. Traditional methods for solving (1) include optimization schemes such as the Gauss–Newton method, or Levenburg–Marquardt method which require derivative information of $\mathcal{G}$, which can prove costly and cumbersome. Therefore a motivation for solving inverse problems is to provide gradient-free optimizers which can reduce this computational burden, while attaining a good level of accuracy. The methodology that we motivate, which alleviates these issues, is that of ensemble Kalman inversion (EKI). EKI can be viewed as the application of the ensemble Kalman filter (EnKF) to inverse problems, which is a natural way to solve inverse problems given the connections between data assimilation and inverse problems. The EnKF is a Monte-Carlo version of the celebrated Kalman filter, which is more favorable in high-dimensions. It operates by updating an ensemble of particles through sample mean and covariances. In particular we will take the viewpoint of EKI which acts as PDE-constrained derivative-free optimizer. Therefore EKI can be viewed as a black-box solver where no derivative information is required. Since this method was proposed for inverse problems, it has seen wide applications to various engineering-based applications, as well as developments related to both theory and methodology. In this chapter we discuss some of these keys concepts and insights, while briefly mentioning particular directions with EKI.

The general outline of these chapter is as follows. In Section 2 we provide the necessary background material, which covers the basics of EKI with some intuition and motivation We will discuss the algorithm in both the usual discrete-time setting, but also the continuous-time setting. This will lead onto Section 3 where we discuss one recent direction which is that of regularization theory, and its application to EKI. Furthermore we will also discuss how EKI can be extended to the notion of sampling in statistics within Section 4. Other, less-developed, directions are provided in Section 5. Numerical experiments are provided in basic settings in Section 6 on a number of basic differential equations, before providing some future remarks and a conclusion in Section 7.

2. EKI: background material

In this section we provide the background material related to the understanding and intiution of EKI. This will begin with a discussion on the ensemble Kalman filter, and how it connections with EKI. We will then present EKI in its vanilla form, which is a discrete-time optimizer, before discussing its connections with various existing methods. Finally we will extend the original formulation to the setting of continuous-time where we aim to provide a gradient flow structure of the resulting equations.

2.1 Kalman filtering

The ensemble Kalman filter (EnKF), is a popular methodology based on the celebrated Kalman filter (KF), which was originally developed by Rudolph Kalman in the 1960s [7, 8]. The Kalman filters initial aim was to solve a recursive estimation problem from dynamics processes and systems. Specifically the KF aims to merge data with model, or signal, dynamics where both equations have the form

$$u_{n+1} = \Psi(u_n) + \xi_n, \qquad \{\xi_n\}_{n\in\mathbb{Z}^+} \sim \mathcal{N}(0,\Sigma), \tag{3}$$

$$y_{n+1} = H(u_{n+1}) + \eta_{n+1}, \qquad \{\eta_{n+1}\}_{n\in\mathbb{Z}^+} \sim \mathcal{N}(0,\Gamma). \tag{4}$$

Here $\{u_n\}_{n\in\mathbb{Z}^+}$ is our signal which is updated through a forward operator $\Psi : \mathbb{R}^m \to \mathbb{R}^m$, which when combined with noise, provides the update u_{n+1}. Our data is denoted as y_{n+1} which is produced by sending our updated signal through the operator $H : \mathbb{R}^m \to \mathbb{R}^{\overline{m}}$, where $\overline{m} > m$, which is known as observational operator. Our initial conditions for the system are given as $u_0 \sim \mathcal{N}(m_0, C_0)$. This area of recursive estimation, in this setup, became to be known as data assimilation [9, 10].

In particular in the linear and Gaussian setting, where the dynamics and noise are Gaussian, the KF updates state using the first two moments, which we know are the mean and covariance. Assume that the state-space dimension is $d \in R^+$, then the cost of the KF has complexity $\mathcal{O}(d^2)$. For high-dimensional examples this can be an issue, therefore an algorithm that was developed to alleviate this is the EnKF, a Monte Carlo version, proposed by Evensen [11, 12].

The EnKF operates by replacing the true covariance by a sample covariance and mean and updates an ensemble of particles $u_n^{(j)}$, with $1 \le j \le J$ particles, using these moments combined with information from the data. The EnKF can be split into a two-step procedure, which is the prediction step

$$\begin{aligned} \hat{u}_{n+1}^{(j)} &= \Psi\left(u_n^{(j)}\right) + \xi_n^{(j)}, \quad \hat{m}_{n+1} = \frac{1}{J}\sum_{j=1}^{J} u_{n+1}^{(j)}, \\ \hat{C}_{n+1} &= \frac{1}{J-1}\sum_{j=1}^{J}\left(u_{n+1}^{(j)} - \hat{m}_{n+1}\right)\left(u_{n+1}^{(j)} - \hat{m}_{n+1}\right)^T, \end{aligned} \tag{5}$$

and update step

$$\begin{aligned} K_{n+1} &= \hat{C}_{n+1}H^T\left(H\hat{C}_{n+1}H^T + \Gamma\right), \\ u_{n+1}^{(j)} &= (I - K_{j+1}H)\hat{u}_{n+1}^{(j)} + K_{n+1}y_{n+1}^{(j)}, \\ y_{n+1}^{(j)} &= y_{n+1} + \eta_{n+1}^{(j)}, \end{aligned} \tag{6}$$

where K_{n+1} represents the Kalman gain matrix and $\xi_n^{(j)}$ and $\eta_{n+1}^{(j)}$ are i.i.d. Gaussian noise. In the EnKF context our prediction step defines a sample mean and covariance from our signal. From this in the analysis step we define our Kalman gain through our sample covariance, which updates our signal, which is given by $u_{n+1}^{(j)}$. This is aided by aiming to minimize the discrepancy of the data $y_{n+1}^{(j)}$ and the quantity $H(u)$. To better understand this discrepancy, there is an alternative approach of looking at the EnKF is through a variational approach, where we consider the follow cost function

$$I_n(u) := \frac{1}{2}\left|y_{n+1}^{(j)} - H(u)\right|_{\Gamma}^2 + \frac{1}{2}\left|u - \hat{u}_{n+1}^{(j)}\right|_{\hat{C}_{n+1}}^2, \tag{7}$$

for which we aim to minimize, which is defined as the updated mean

$$\hat{m}_{n+1} = \arg\min_{u} I_n(u). \tag{8}$$

This minimization procedure relies on the updated covariance $\hat{C}_{n+1}$ which is dependent entirely on $\hat{u}^{(j)}$. As described in the prediction step and update step of filtering, a mapping is presented between distributions. As we related the distributions in the filtering setting, for each step, we can do so similarly for the EnKF, i.e.

$$\left\{u_n^{(j)}\right\}_{j=1}^{J} \mapsto \left\{u_{n+1}^{(j)}\right\}_{j=1}^{J}, \qquad \left\{u_{n+1}^{(j)}\right\}_{j=1}^{J} \mapsto \left\{\hat{u}_{n+1}^{(j)}\right\}_{j=1}^{J}. \tag{9}$$

With the EnKF, compared to KF, the computational complexity associated with it is $\mathcal{O}(Jd)$, where one usually assumes $J < d$, therefore implying the reduction in cost.

2.2 EnKF applied to inverse problems

Since the formulation of the EnKF, there has been a huge interest from practitioners in various applicable disciplines. Most notably this has been within numerical weather prediction, geophysical sciences and signal processing related to state estimation. In this chapter our focus is on the application of the EnKF to inverse problems, namely to solve (1). We now introduce this application which is known as ensemble Kalman inversion (EKI), which was introduced by Iglesias et al., motivated from Li et al., [13] as a derivative-free optimizer for PDE-constrained inverse problems.

As with the EnKF, we are concerned with updating an ensemble of particles, for which now we modify notation with n now denoting the iteration count. Given an initial ensemble $\left\{u_0^{(j)}\right\}$, our aim is to learn a true underlying unknown $u^\dagger$. To do so, as done with the EnKF, we first define our sample mean and covariance matrices

$$\overline{u}_n^{(j)} = \frac{1}{J}\sum_{j=1}^{J} u_n^{(j)}, \quad \overline{u}_n^{(j)} = \frac{1}{J}\sum_{j=1}^{J} G\left(u_n^{(j)}\right),$$

$$C_n^{uu} = \frac{1}{J-1}\sum_{j=1}^{J}\left(u_n^{(j)} - \overline{u}\right)\left(u_n^{(j)} - \overline{u}\right)^T, \quad C_n^{up} = \frac{1}{J-1}\sum\left(u_n^{(j)} - \overline{u}\right)\left(G\left(u_n^{(j)}\right) - \overline{G}\right)^T. \tag{10}$$

which we can through the update equation

$$u_{n+1}^{(j)} = u_n^{(j)} + hC^{up}\left(hC^{pp} + \Gamma\right)^{-1}\left(y_n^{(j)} - \mathcal{G}\left(u_n^{(j)}\right)\right), \tag{11}$$

$$y_n^{(j)} = y + \eta_n^{(j)}, \tag{12}$$

where y represents our true data and $h > 0$ denotes a step size related to the level of discretization. **Figure 1** provides a pictorial description of the EnKF, which has been described above.

The update equation of EKI (11) is of interest as it coincides with the update formula for Tikhonov regularization for linear statistical inverse problems. Namely if we consider $R(u) = \frac{1}{2}\|u\|_{C_0}^2$, then the update formula, in the linear $\mathcal{G}(\cdot) = \mathcal{G}$ and Gaussian setting is given as

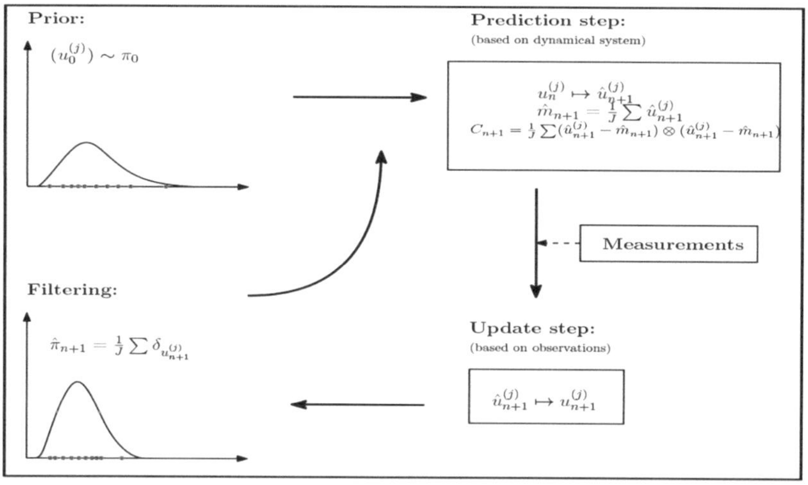

Figure 1.
Dynamics of the ensemble Kalman filter, split into the prediction and update steps.

$$u_{TP} = \overline{u} + C\mathcal{G}^* \left(\mathcal{G}C\mathcal{G}^* + \Gamma\right)^{-1}(y - \mathcal{G}\overline{u}), \tag{13}$$

where $\mathcal{G}^*$ denotes the derivative of the operator $\mathcal{G}$. This connection is of relevance and was discussed in [14], where it was shown that taking the limit as $J \rightarrow \infty$, it was shown that $u \rightarrow u_{TP}$. This is of interest as the minimizing the regularized functional (13) is equivalent to the following maximization procedure in statistics

$$u{:=}\arg\max_{u \in \mathcal{X}} \mathbb{P}(u|y). \tag{14}$$

known as the MAP formulation, where $\mathbb{P}(u|y) = \mathbb{P}(y|u)\mathbb{P}(u)$ denotes the posterior distribution. This connection is discussed in [15]. Therefore this provides some insight into EKI and its connection with other known existing methodologies in inverse problems. An important entity to discuss is a property that EKI inherits, which is the *subspace property*. It is given by the following lemma.

Lemma 1.1 Let $\mathcal{A}$ be the linear span of the initial ensemble $\left\{u_0^{(j)}\right\}_{j=1}^J$, then we that $\left\{blacku_n^{(j)}\right\}_{j=1}^J \in \mathcal{A}$ for all $n \in \mathbb{N}$.

The essence of the subspace property states that the updated ensemble of particles is spanned by the initial ensemble. This is important, because it provides a justification on the performance, whether the initial ensemble is a good choice or not. Therefore it can act as an advantage or a disadvantage.

2.3 Continuous-time formulation

The original representation of EKI, as shown in (11), is a discrete-time iterative scheme similar to other optimization methods. However it is of interest to understand EKI in a continuos-time setting, which was considered by Schillings et al. [16, 17]. This is primarily for two reasons; (i) firstly that one can understand more easily how the dynamics of (11) and (12) behaves, and secondly (ii) it provides new numerical schemes for EKI, which is specific in the continuous-time setting. In order to derive such equations, as usual we require to take the step-size to zero, i.e. $h \to 0$. Once we do this, we have the following set of stochastic differential equations

$$\frac{du^{(j)}}{dt} = C^{uw}(u)\Gamma^{-1}\left(y - \mathcal{G}\left(u^{(j)}\right)\right) + C^{uw}(u)\sqrt{\Gamma^{-1}}\frac{dW^{(j)}}{dt}, \tag{15}$$

with $W^{(j)}$ denoting independent cylindrical Brownian motions. By substituting the form of the covariance operator, we see

$$\frac{du^{(j)}}{dt} = \frac{1}{J}\sum_{k=1}^{J}\left\langle \mathcal{G}\left(u^{(k)}\right) - \overline{\mathcal{G}}, y - \mathcal{G}\left(u^{(j)}\right) + \sqrt{\Gamma}\frac{dW^{(j)}}{dt}\right\rangle_{\Gamma}\left(u^{(k)} - \overline{u}\right). \tag{16}$$

For this we take our forward operator $\mathcal{G}(\cdot) = A\cdot$ to be bounded and linear. Using this notion and by substituting our linear operator A in (16) we have the following diffusion limit

$$\frac{du^{(j)}}{dt} = \frac{1}{J}\sum_{k=1}^{J}\left\langle A\left(u^{(k)} - \overline{u}\right), y - Au^{(j)}\right\rangle_{\Gamma}\left(u^{(k)} - \overline{u}\right). \tag{17}$$

By defining the empirical covariance operator

$$C(u) = \frac{1}{J-1}\sum_{k=1}^{J}\left(u^{(k)} - \overline{u}\right) \otimes \left(u^{(k)} - \overline{u}\right), \tag{18}$$

and taking $\Gamma = 0$ we can express (17) as

$$\begin{aligned}\frac{du^{(j)}}{dt} &= -C(u)D_u\Phi\left(u^{(j)};y\right),\\ \Phi(u;y) &= \frac{1}{2}\|\Gamma^{-1/2}(y - Au)\|^2.\end{aligned} \tag{19}$$

Thus we note that each particle performs a preconditioned gradient descent for $\Phi(\cdot;y)$ where all the gradient descents are preconditioned through the covariance $C(u)$. Since our covariance operator $C(u)$ is semi-positive definite we have that

$$\frac{d}{dt}\Phi(u(t);y) = \frac{d}{dt}\frac{1}{2}\|\Gamma^{-1/2}(y - Au)\|^2 \leq 0. \tag{20}$$

In the context of EKI this is of interest as it is a first result providing some indication of the dynamics, which was not achievable through the discrete-time update formula (11). Indeed given the gradient flow structure, we are able to see that the EKI abides by a usual optimization function, with the dynamics following the direction of the negative gradient, or in other-words towards to minimizer of Φ. Since the continuous-time formulation was derived, there has been different works deriving further analysis, most notably with recent success on the nonlinear setting, and other well-known results. This can be found in [18].

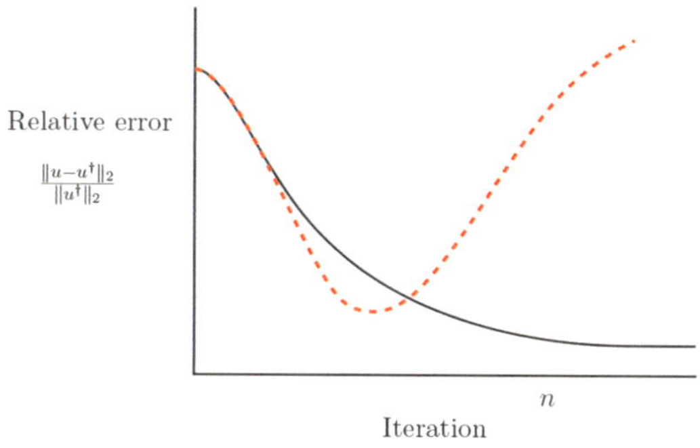

Figure 2.
The figure presents two simulations of EKI as the iterations increase. The black curve represents what we aim to achieve, however in certain situations the data is commonly overfitted. Therefore this can cause a divergence in the relative error, as shown by the dashed red curve.

3. Regularization

In this section we discuss the role of regularization in EKI. We will begin with an introduction into iterative regularization schemes, that have been used before discussing Tikhonov regularization, L_p and particular adaptive choices.

As briefly discussed regularization is an important tool in optimization, and inverse problems aimed at preventing the over-fitting, or influence, of the data. We refer the reader to various pieces of literature that give a concise overview on this [19, 20]. The over-fitting of data can cause issues in inverse problems, such as the divergence of the error, therefore careful consideration is needed to prevent this. A cartoon representation of this is given in **Figure 2**.

To initiate this chapter, there are two main forms of regularization one can apply for inverse problems. The first is related to *iterative regularization*, where the regularization is included within the iterative scheme. This can be included directly such as the form

$$u_{n+1}^{(j)} = u_n^{(j)} + hC^{up}(hC^{pp} + \alpha_n\Gamma)^{-1}\left(y_n^{(j)} - \mathcal{G}\left(u_n^{(j)}\right)\right), \tag{21}$$

or in the presence of a discrepancy principle of the form

$$\|\Gamma^{-1}\left(y - \overline{u}_n^{(j)}\right)\|^2 \le \vartheta\eta, \qquad \vartheta \in (0, 1), \tag{22}$$

which controls the error between the updated ensemble and the true unknown. The discrepancy principle acts as a stopping rule if the error becomes big, and the the modified update formula contains a sequence of numbers $\{\alpha_n\}_{n\in\mathbb{N}}$ aimed at also preventing the overfitting of the data. This sequence is chosen in such a way that is related to a discrepancy principle. Specifically for EKI this has been considered in numerous work by Iglesias et al. [21, 22].

However more recent work has considered regularization through the least-squares functional (LSF) (2). For EKI the first known form to consider this, is Tikhonov regularization which has the penalty form of $R(u) = \frac{1}{2}\|u\|_{C_0}^2$. This form of regularization is a natural choice, as it very well-known and understood but can view viewed as a Gaussian form of regularization, which smoothes the problems. In the context of EKI this makes sense, as commonly one assumes Gaussian dynamics. The work of Chada et al. [23] first developed this extension, which was done by modifying (1) to the following

$$\begin{aligned} y &= G(u) + \eta_1, \\ u &= \eta_2, \end{aligned} \tag{23}$$

where $\eta_1 \sim \mathcal{N}(0, \Gamma)$,$\eta_2 \sim \mathcal{N}(0, \lambda^{-1}C_0)$..
Now we introduce z,η and the mapping $\mathcal{F} : \mathcal{X} \times \mathcal{X} \mapsto \mathcal{Y} \times \mathcal{X}$ as follows:

$$z = \begin{bmatrix} y \\ 0 \end{bmatrix}, \quad F(u) = \begin{bmatrix} G(u) \\ u \end{bmatrix}, \quad \eta = \begin{bmatrix} \eta_1 \\ \eta_2 \end{bmatrix}, \tag{24}$$

and

$$\eta \sim N(0, \Sigma), \quad \Sigma = \begin{bmatrix} \Gamma & 0 \\ 0 & \lambda^{-1}C_0 \end{bmatrix}. \tag{25}$$

Therefore our inverse problem is now reformulated at

$$z = \mathcal{F}(u) + \eta. \tag{26}$$

now from this we can modify EKI to include the above setup, for which we refer to it as Tikhonov ensemble Kalman inversion (TEKI), which takes the following form

$$u_{n+1}^{(j)} = u_n^{(j)} + hB^{up}(hB^{pp} + \Gamma)^{-1}\left(z_n^{(j)} - \mathcal{F}\left(u_n^{(j)}\right)\right), \tag{27}$$

where we have now modified covariance matrices B^{up},B^{pp}. From this inclusion, the authors of [23] were able to show that analytically, the subspace property still holds, while other such results as observability and controllability and the ensemble collapse. More importantly through the numerical simulations, it was shown that one can prevent the over-fitting phenomenon.

Since this work a number of useful extensions have been considered, such as its understanding in the continuous-case, as well as the new variants in the discrete-time setting [24]. Two recent developments on this have been firstly on the extension to L_p regularization [25, 26], which is to motivate reconstructing edges or lines, where the LSF is modified to

$$\Phi(u;y) := \frac{1}{2}\|y - \mathcal{G}(u)\|_\Gamma^2 + \lambda\|u\|_p, \quad p \geq 1. \tag{28}$$

Finally another direction is related to producing adaptive strategies for TEKI. Adaptive regularization schemes are of importance, as choosing a correct choice of the regularization parameter $\lambda > 0$ can have a big impact on the reconstruction. Therefore thinking adaptively allows one to evolve the parameter over the iteration count, now denoted as λ_n. The work of Weissmann et al. [27] provides these developments in an adaptive fashion.

4. Ensemble Kalman sampling

Although the EKI has been introduced through the application of the EnKF to inverse problems and hence sequential sampling method, the trending viewpoint of

EKI lies in optimization. So far, we have seen its motivation from the gradient flow structure in the continuous-time formulation in Section 2.3 and the representation as SDE. For applying EKI as a consistent sampling method, we would instead of taking the limit $t \to \infty$ rather consider the limit $t \to 1$. For linear forward models EKI is consistent with the posterior distribution, however, it is known to be not consistent with the Bayesian perspective in the nonlinear setting [28].

Building up on this fact, the motivation behind the ensemble Kalman sampler [29] is to modify the time-dynamical system of EKI in a way such that the limiting distribution for $t \to \infty$ corresponds to the posterior distribution. We will start the discussion with an introductory example.

Example 1.1 Let π_* be a pdf of the form $\pi_*(x) \propto \exp(-\Phi(u))$ with $\Phi(u) = \frac{1}{2}\|y - G(u)\|_\Gamma^2 + \|u\|_C^2$, i.e. π_* corresponds to the posterior pdf under Gaussian prior assumption $\pi_0 = \mathcal{N}(0, C)$. We consider the Langevin diffusion given by

$$du_t = \nabla_u \log \pi_*(u_t)dt + \sqrt{2}dW_t, \quad u_0 \sim \pi_0, \tag{29}$$

where $(W_t)_{t\geq 0}$ denotes a Brownian motion in $\mathcal{X} = \mathbb{R}^{n_u}$. The evolution of the distribution ρ_t of the state u_t can then be described through the Fokker–Planck equation

$$\partial\rho_t = \nabla \cdot (\rho_t \nabla \log \pi_*) + \Delta\rho_t, \quad \rho_0 = \pi_0, \tag{30}$$

where under certain assumptions on Φ the underlying Markov process $(u_t)_{t\geq 0}$ is ergodic and its unique invariant distribution is given by π_* [30]. Taking the Fokker–Planck eq. (30) into account the convergence to equilibrium can be described through the Kullback–Leibler (KL) divergence $\mathrm{KL} = \int_{\mathcal{X}} q_1(x) \log\left(\frac{q_1(x)}{q_2(x)}\right) \mathrm{d}x$ [31]. Assuming a log-Sobolev inequality (e.g. satisfied for log-concave π_*), it follows that

$$\mathrm{KL}(\rho_t|\pi_*) \leq \exp(\ \lambda t)\mathrm{KL}(\rho_0|\pi_*) \tag{31}$$

for some $\lambda > 0$ [32].

4.1 Interacting Langevin sampler

The interacting Langevin sampler has been introduced, motivated by the preconditioned gradient descent method, as interacting particle system represented by the coupled system of SDEs

$$du_t^{(j)} = C(u_t)\nabla_u \log \pi_*\left(u_t^{(j)}\right)dt + \sqrt{2C(u_t)}dW_t, \quad j = 1, \dots, J, \tag{32}$$

initialized through an i.i.d. sample $u_0^{(j)} \sim \pi_0$. The idea of preconditioning with $C(u_t)$ instead of a fixed preconditioning matrix $C \in \mathbb{R}^{n_u \times n_u}$ is motivated through the corresponding mean-field limit. In the large particle limit, the corresponding SDE is given as

$$du_t = C(\rho_t)\nabla_u \log \pi_*(u_t)dt + \sqrt{2C(\rho_t)}dW_t, \quad u_0 \sim \pi_0, \tag{33}$$

where the macroscopic mean and covariance operator are defined as

$$m(\rho) = \int_{\mathcal{X}} x\rho(x)dx, \quad C(\rho) = \int_{\mathcal{X}} (x - m(\rho)) \otimes (x - m(\rho))dx. \tag{34}$$

This connects the interacting Langevin system to its origin Langevin diffusion (29). Hence, in the long-time limit the preconditioning matrix will formally be given by the covariance operator corresponding to the stationary distribution (assuming it exists).

The resulting modified Fokker–Planck equation is given by

$$\partial\rho_t = \nabla \cdot (\rho_t C(\rho_t)\nabla \log \pi_*) + \mathrm{Tr}\big(C(\rho_t)D^2\rho_t\big), \quad \rho_0 = \pi_0. \tag{35}$$

Assuming that $C(\rho_t) \geq \alpha \mathrm{Id}$ and the target distribution of the form $\pi_*(u) \propto \exp(-\Phi(u))$, $\Phi(u) = \frac{1}{2}\|y - \mathcal{G}(u)\|_\Gamma^2 + \lambda\|u\|_{C_0}^2$, to be $\log$-concave, the solution ρ_t of (35) converges exponentially fast to equilibrium

$$\mathrm{KL}(\rho_t|\pi_*) \leq \exp(-\lambda t)\mathrm{KL}(\rho_0|\pi_*), \tag{36}$$

for some $\lambda > 0$ [29], Proposition 3.1. Furthermore, through the preconditioning with the sample covariance the resulting scheme remains invariant under affine transformations [33].

4.2 Ensemble Kalman sampler

One of the attractive features of the EnKF as well as of EKI is its derivative-free implementation. The basis of the ensemble Kalman sampler (EKS) is to build a modified interacting Langevin sampler avoiding to compute derivatives. Let $\pi_*(u) \propto \exp\left(-\frac{1}{2}\|y - \mathcal{G}(u)\|_\Gamma^2 - \|u\|_{C_0}^2\right)$, then the interacting Langevin system is given by

$$du_t^{(j)} = -C(u_t)D\mathcal{G}\left(u_t^{(j)}\right)^{\mathrm{T}}\Gamma^{-1}\left(G\left(u_t^{(j)}\right) - y\right) - C(u_t)C_0^{-1}u_t^{(j)}dt + \sqrt{2C(u_t)}dW_t, \quad j = 1, \dots, J. \tag{37}$$

Motivated by the approximation $C^{uw}(u) \approx C(u)D\mathcal{G}\big(u^{(j)}\big)^{\mathrm{T}}$ the EKS is then formulated as the solution of the system of coupled SDEs

$$du_t^{(j)} = -C^{uw}(u_t)\Gamma^{-1}\left(\mathcal{G}\left(u_t^{(j)}\right) - y\right) - C(u_t)C_0^{-1}u_t^{(j)}dt + \sqrt{2C(u_t)}dW_t, \quad j = 1, \dots, J. \tag{38}$$

We note that the approximation $C^{uw}(u) \approx C(u)D\mathcal{G}\big(u^{(j)}\big)^{\mathrm{T}}$ is exact for linear forward models and hence, the EKS coincides with the interacting Langevin sampler in the linear setting. However, for nonlinear forward models the approximation of derivatives is only accurate in case the particles are close to each other. Since in the application of EKS the particles are aiming to represent a distribution, the particles are not expected to be close to each other. This fact suggests to formulate a localized version of the preconditioning sample covariance matrix, incorporating more weights on particles close to each other, but reducing the weight between particles far away. Therefore, we define the distance-dependent weights between particle $u_t^{(j)}$ and $u_t^{(i)}$

$$w_t^{ji} = \frac{\exp\left(-\frac{1}{2\gamma}\|u_t^{(j)} - u_t^{(i)}\|_D^2\right)}{\sum_{l=1}^J \exp\left(-\frac{1}{2\gamma}\|u_t^{(j)} - u_t^{(l)}\|_D^2\right)}, \tag{39}$$

for scaling parameters $\gamma > 0$ and symmetric positive-definite matrix $D \in \mathbb{R}^{n_u \times n_u}$. The localized (mixed) sample covariance matrix around particle $u_t^{(j)}$ is the defined as

$$\begin{aligned} C\left(u_t^{(j)}\right) &= \sum_{i=1}^J w_t^{ji}\left(u_t^{(i)} - \overline{u}_t^{(j)}\right) \otimes \left(u_t^{(i)} - \overline{u}_t^{(j)}\right), \\ C^{uw}\left(u_t^{(j)}\right) &= \sum_{i=1}^J w_t^{ji}\left(u_t^{(i)} - \overline{u}_t^{(j)}\right) \otimes \left(\mathcal{G}\left(u_t^{(i)}\right) - \overline{\mathcal{G}}_t^{(j)}\right), \end{aligned} \tag{40}$$

with localized mean

$$\overline{u}_t^{(j)} = \sum_{i=1}^J w_t^{ji} u_t^{(i)}, \quad \overline{\mathcal{G}}_t^{(j)} = \sum_{i=1}^J w_t^{ji} \mathcal{G}\left(u_t^{(i)}\right). \tag{41}$$

The localized EKS then reads as

$$du_t^{(j)} = -C^{uw}\left(u_t^{(j)}\right)\Gamma^{-1}\left(\mathcal{G}\left(u_t^{(j)}\right) - y\right) - C\left(u_t^{(j)}\right)C_0^{-1}u_t^{(j)}\,dt + \sqrt{2C\left(u_t^{(j)}\right)}dW_t, \quad j = 1, \dots, J. \tag{42}$$

While the original EKS shows promising results for nearly Gaussian target distribution, the considered localized variant helps to extend the scope to multimodal target distributions [34]. Other such related work has aimed to provide further understandings of the EKS. This has included the derivation of providing mean field limits and, but also providing various generalizations [33, 35].

5. Other directions

As we have discussed some of the more recent developments in EKI, we now focus on other, more smaller, extensions. In this section we will discuss these each in turn, which will include machine learning, understanding EKI in the context of nonlinear inverse problems, and finally applications related to engineering such as geophysical sciences.

5.1 Applications in machine learning

The developments of machine learning methodologies has seen a significant increase in the last decade, which have been produced to solve problems related to health-care, imaging, and decision processes. In particular much of the these developments has been to due the advancements in optimizaion theory. As a result, ensemble Kalman methods can be viewed as a natural class of algorithms to be directly applied, as they are derivative-free optimizers.

The first work aimed at characterizing this connection was [36] which demonstrated this. The authors motivated EKI as a replacement to SGD where they initially applied it to supervising learning problems. Given a dataset $\{x_j, y_j\}_{j=1}^{N}$ assumed to be i.i.d. samples from a particular distribution, then given the Monte Carlo approximation one has the minimization procedure

$$\arg\min_{u} \Phi_s(u; x, y),$$
$$\Phi_s(u; x, y) = \frac{1}{N}\sum_{i=1}^{N} \mathcal{L}\left(\mathcal{G}\left(u|x_j\right), y_j\right) + \frac{\lambda}{2}\|u\|_{C_0}^2, \tag{43}$$

where $\mathcal{L} : \mathcal{Y} \times \mathcal{Y} \to \mathbb{R}^+$, is some positive-definite function. In other words, one is trying to learn x_j from the labeled data y_j. Supervised learning is used for common ML applications such as image classification and natural language processing. Another related application is that of semi-supervised learning, which aims to learn x_j from some of the data y_j where do not have access to all of it. This modified the least squares functional given in (43).

Another interesting direction has been the inclusion of EKI for training and learning neural networks [37]. This builds upon the previous work discussed, but with a number of modifications. In particular what the authors show is that they are able to prove convergence of EKI to the minimizer of a strongly convex function. They apply their modified methodology to a nonlinear regression problem of the form

$$F(\theta) = A\theta + \varepsilon \sin(B\theta), \tag{44}$$

where θ is the parameter of interest and $F(\theta)$ is the objective functional of interest. This was also extended to the likes of image classification problems, specifically the well-known MNIST handwritten data set.

A final and more recent direction of EKI and ML, was the work of Guth et al. [38], which provided a way of solving the forward problem, within EKI.

5.2 Extensions to nonlinear convergence analysis

A major challenge with EKI, and the EnKF in general, is establishing convergence analysis and properties in the nonlinear setting. As it is well known in the linear and Gaussian setting, as the the number of particles $N \to \infty$, the EnKF coincides with the KBF. However in the nonlinear setting it is has been challenging to derive any such results rigorously. Some ongoing and recent work has aimed to bridge the connections between EKI and nonlinear dynamics. The first paper that provided some form of analysis was the work of Chada et al. [24] which considered a specific form of EKI, in the discrete-time setting.

Namely the update formula is modified to

$$m_{n+1} = m_n + C_n^{pp}\left(C_n^{up} + h_n^{-1}\Gamma\right)^{-1}(z - H(m_n)),$$
$$C_{n+1} = C_n^{uu} - C_n^{up}\left(C_n^{pp} + h_n^{-1}\Gamma\right)^{-1}C_n^{pu} + \alpha_n^2\Sigma, \tag{45}$$

where we adopt an ensemble square root filter formulation, which is known to perform better. As well as this we also include covariance inflation (i.e. inflation factor of α_n), and an adaptive step-size h_n motivated from stochastic optimization to allow an acceleration for the convergence. However the other underlying contribution, as eluded to, is that given this update form we are able to prove convergence towards both local and global minimizers. In other words for the later, we have the following result

$$\lambda_c \|m_N - u^*\|^2 \leq \ell(m_N) - \ell(u^*) \leq \frac{D}{N^\alpha}, \tag{46}$$

which the above result establishes polynomial convergence. We note from the above equation, that λ_c is a convexity constant, ℓ is the associated loss function, D is some constant, u^* is the global minimizer and α is some term, which we refer to [24], for further details.

As one can notice, this convergence analysis was considered for the discrete-time setting, so a natural extension from this is to the continuous-time framework. The work of Blomker et al. [18] provide a first convergence analysis in this direction. However given both these works, a full understanding in the nonlinear setting has not been achieved, where considerable work is still required. Thus these papers provide a first step in doing so, for both settings.

5.3 Engineering applications

As a final direction to discuss in detail, which is very much related to the theme of this book, are applications in particular engineering applications. The advantage of these ensemble Kalman methods, is that they can be viewed as a black box-solver, therefore it is highly applicable. One particular application has been geophysical sciences, related to recovering quantities of interest which are below the surface, or subsurface. Examples include the inverse problem of electrical resistivity tomography (ERT), shown below (**Figure 3**).

ERT is concerned with recovering, or characterizing sub-surface materials in terms of their electrical properties, which are recorded through electrodes. It operates very similarly to electrical impedance tomography (EIT), expect the difference being that it is subsurface. This has been also considered for learning permeability of subsurface flow in a range of different settings which can be found in the following papers [39, 40].

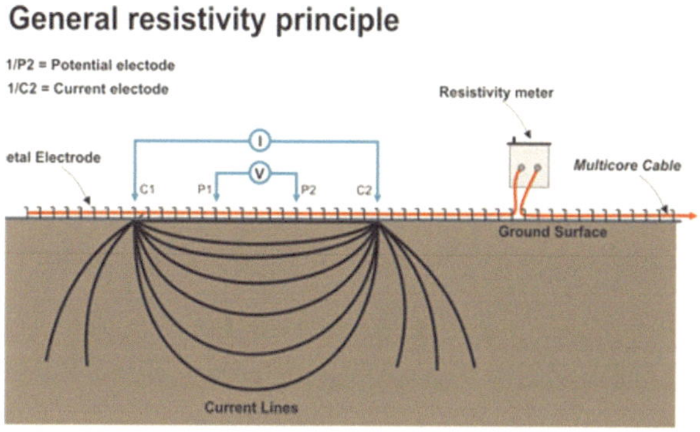

Figure 3.
Image depicting electrical resistivity tomography, where the the electric currents are recorded at the electrodes of the subsurface material.

Another interesting direction is related to walls, specifically quantifying uncertainty in thermo-physical properties of walls. This work was conducted by Iglesias et al. [41, 42]. Specifically the application is the inverse problem of recovering the thermodynamic property or temperature. Similar work related to the methodology used here has been used in resin transfer modeling [43], based on problems of moving boundaries. This is a difficult problem to model, however it provides a first step in doing so. Aside from these applications other particular applications include mineral exploration scattering problems, numerical climate models and others [44–46]. It is worth mentioning that, as of now, there is no official online software package for EKI in general. This is currently being developed, but we emphasize to the reader that the methodology presented, with the examples later, are not related to well known softwares that are available in `Matlab` or `Python`.

As a side remark, there are more directions beyond what is discussed above. Some others, without going into details, include developing hierarchical approaches, incorporating constrained optimization, and connections with data assimilation strategies [47–51].

6. Numerical experiments

In this section we provide some numerical experiments highlighting the performance of ensemble Kalman methods for inverse problems. Specifically we will consider EKI as discussed in Section 2. We will compare EKI with its regularized version of TEKI. Both these methodologies will be tested on on two motivating inverse problems arising in geophysical and atmospheric sciences, i.e. a Darcy flow partial differential equation and the Navier–Stokes Equation.

In order to assess a comparison, we will present three different figures. (i) The first being a reconstruction at the end of the iterative scheme; (ii) the error between the approximate solution and the ground truth, and (iii) the data misfit. The equations associated with each are given as.

- *Reconstruction through EKI:* $\frac{1}{J}\sum_{j=1}^{J} u_n^{(j)}$.
- *Relative error:* $\frac{\|u^\dagger - u\|_{L^2}^2}{\|u^\dagger\|_{L^2}}$.
- *Data misfit:* $\|\Gamma^{-1/2}(y - \mathcal{G}(u^\dagger)\|^2$.

6.1 Darcy flow

Our first model problem is an elliptic partial differential equation (PDE), which has numerous applications. Specifically one of them is subsurface flow in a porous medium. The forward problem is concerned with solving for the pressure $p \in H_0^1(\Omega)$, given the permeability $\kappa \in L^\infty(\Omega)$ and source function $f \in L^\infty(\Omega)$, where the PDE is given as

$$-\nabla \cdot (\kappa \nabla p) = f, \quad \in \Omega, \tag{47}$$

$$p = 0, \quad \text{on } \Omega. \tag{48}$$

such that we have prescribed Dirichlet boundary conditions, and $\Omega = [0, 1]^2 \subset \mathbb{R}^d$, for $d = 2$, is a Lipschitz domain. The inverse problem associated to solving p from (47) is the recovery of the permeability $\kappa \in L^\infty(\Omega)$, from noisy measurements of p, i.e.

$$y = \mathcal{G}(\kappa) + \eta, \quad \eta \sim \mathcal{N}(0, \Gamma), \tag{49}$$

where recalling that $\mathcal{G}(\kappa) = p$. We consider 64 equidistance observations within the domain, and on the boundary. To numerically solve (47) we employ a centered-finite difference method with a mesh size of $h = 1/100$. For our noisy observations we consider $\Gamma = \gamma I$, where $\gamma = 0.01$. We will use and compare EKI and TEKI, with an ensemble size of $J = 50$ for both methods. We will run both iterative schemes for $n = 24$ iterations. For our initial ensemble $\{u_0\}_{j=1}^{J}$ we consider modeling it as a Gaussian random field, i.e. $u \sim N(0, C)$, which can be done via the Karhunen-Loève expansion

$$u = \sum_{k \in \mathbb{Z}^+} \sqrt{\lambda_k} \phi_k \xi_k, \quad \xi_k \sim N(0, 1), \tag{50}$$

where (λ_k, ϕ_k) are the associated eigenvalues and eigenvectors of the covariance operator C. There are numerous choice of covariance functions one can take, however a popular choice is the Matérn covariance function, which provides much flexibility for modeling. For full details on various covariance functions, or operators, we refer to reader to [52]. The true unknown of interest is taken to be also a Gaussian random field, but one that is smoother than that of that of the initial ensemble.

Our first set of experiments are provided in **Figure 4** which shows the truth, the reconstruction from using EKI, and that of using TEKI. As we can observe, is it clear that both methodologies work well at learning the true unknown function. However it is clear that the TEKI induces a smoother reconstruction, which arises from the regularization. However, what is interesting is that if we analyze **Figure 5**, we notice

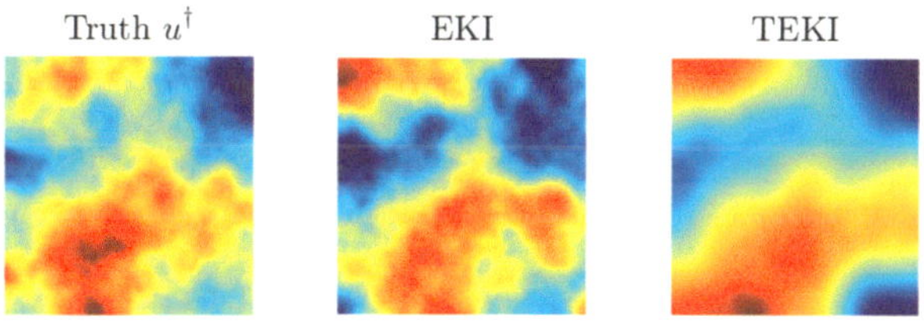

Figure 4.
Reconstruction plots for the Darcy flow PDE example. Left: Truth. Middle: EKI reconstruction. Right: TEKI reconstruction.

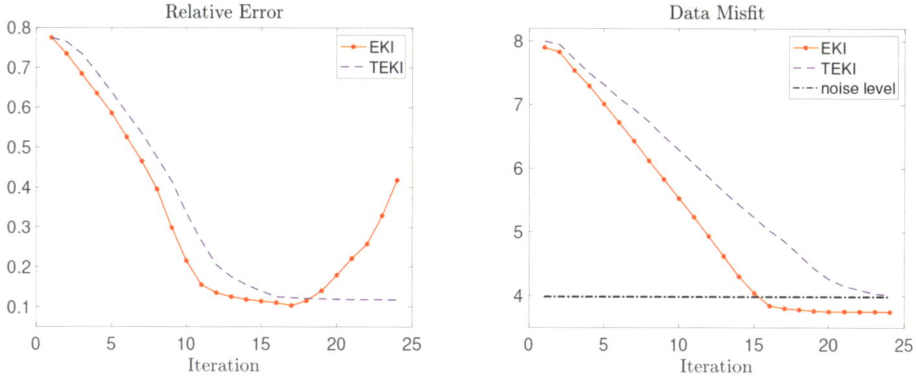

Figure 5.
Relative errors and data misfits for the Darcy flow PDE example. We compare EKI with TEKI.

that the relative error tends to diverge at the end with EKI, and this is due to the overfitting of data. A motivation behind TEKI is to alleviate this. This can be seen vividly as it tends to decrease, and for the data misfit, it remains within the noise level, which is given as

$$\text{noise level} = \|(y - G(u^\dagger))\| = \|\eta^\dagger\|. \tag{51}$$

6.2 Navier: stokes equation

Our final test problem is a well-known PDE model arising in numerical weather prediction which is the Navier–Stokes equation (NSE). We consider a 2D NSE defined on a torus $\mathbb{T}^2 = [0, 1]^2$ with periodic boundary conditions. The aim to estimate the velocity $v{:=}[0, \infty) \times \mathbb{T}^2 \to \mathbb{R}^2$ defined as a vector field from the scalar pressure field $p{:=}[0, \infty) \times \mathbb{T}^2 \to \mathbb{R}^2$. The NSE is given as

$$\partial_t v + (v \cdot \nabla)v + \nabla p - \nu \Delta v = f, \quad [0, \infty) \times \mathbb{T}^2, \tag{52}$$

$$\nabla \cdot v = 0, \quad [0, \infty) \times \mathbb{T}^2, \tag{53}$$

$$v = u, \quad \{0\} \times \mathbb{T}^2, \tag{54}$$

with initial condition (54) and zero flux (53). From (52) $f \in [0, \infty) \times \mathbb{T}$ corresponds to a volume forcing, ν is the associated viscosity of the fluid. For the NSE equation we consider a spectral Fourier solver for (52). The PDE is more challenging

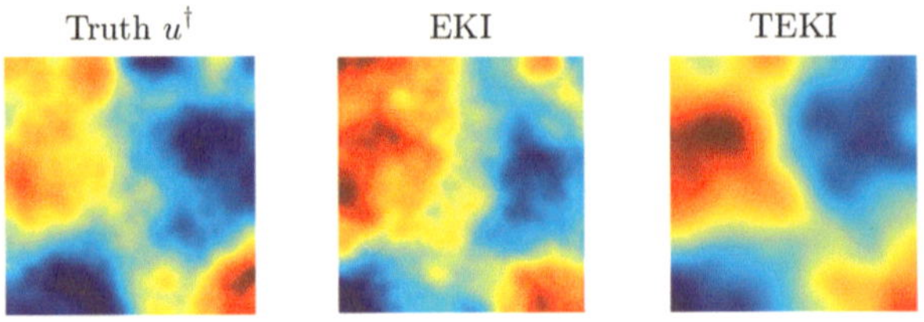

Figure 6.
Reconstruction plots for the NSE PDE example. Left: Truth. Middle: EKI reconstruction. Right: TEKI reconstruction.

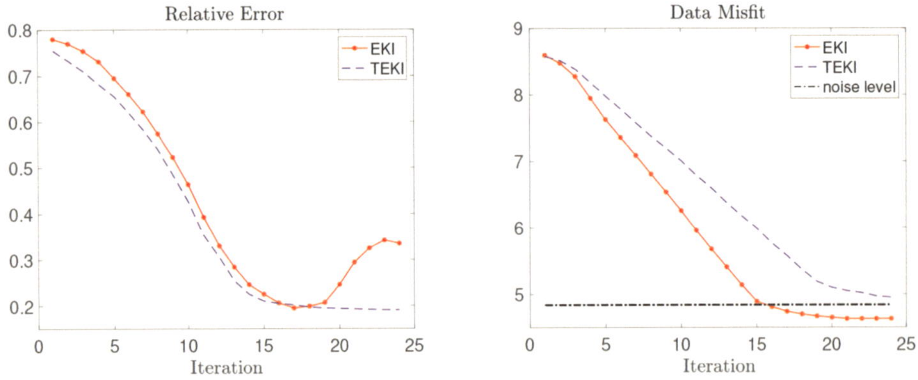

Figure 7.
Relative errors and data misfits for the NSE PDE example. We compare EKI with TEKI.

to invert than the previous example, therefore we take 100 point-wise observations. The setup is largely the same as the previous example, where we take an initial condition based on a Gaussian random field through the KL expansion (50). We will aim to recover the true underlying function $u^\dagger$ using both EKI and TEKI. The results are obtained from the experiments are presented in **Figures 6** and 7. A similar phenomenon shows, where the reconstructions work well, however there is an additional smoothness induced through the regularization in TEKI. Similarly, as we see with the relative errors and data misfit the overfitting of the data in the end for EKI. We note that this can be avoided depending on the prior form, its hyperparameters, the observations, and the noise. However we specify particular choices to demonstrate it can occur.

7. Conclusion

The ensemble Kalman filter (EnKF) is a simplistic, easy-to-implement and powerful algorithm. This has been particularly the case in numerous data assimilation applications for state estimation, which includes the likes of numerical weather prediction, geosciences and more recently machine learning. A major advantage of the method is that, unlike other filters such as the particle filter, it scales better in high dimensions, and can be significantly cheaper. In this chapter we consider the EnKF and its application to parameter estimation. Such a mathematical procedure also has similar applications to the ones states, where one can exploit such techniques for inverse problems. We provide a review and overview of some of the major contributions in this direction, where the resulting methodology is known as ensemble Kalman inversion (EKI), based largely on the work of Iglesias et al. [13]. We presented various avenues the field of EKI has taken such as regularization, extensions to sampling, and other areas. We demonstrated how EKI can perform on two numerical examples PDE examples.

The EKI methodology is one which builds very naturally from many different fields, which acts a strong motivation. For example being an optimizer, one can naturally apply optimization procedures, but also techniques from data assimilation and uncertainty quantification. As a result, this methodology naturally brings researchers from different fields working towards parameter estimation, and inverse problems. This synergy of areas will hopefully ensure new emerging directions within EKI, from a methodological, theoretical and application perspective.

Acknowledgements

This work was funded by KAUST baseline funding. The author thanks Simon Weissmann for helpful discussions, and for the use of some of the earlier figures, and information on EKS.

Abbreviations

EnKF	Ensemble Kalman filter
EKI	Ensemble Kalman inversion

EKS	Ensemble Kalman sampler
EIT	Electrical impedance tomography
ERT	Electrical restivitiy tomography
MNIST	Modified National Institute of Standards and Technology
KL	Kullback-Leibler
KF	Kalman filter
L96	Lorenz 96 model
LSF	Least squares functional
PDE	Partial differential equation
SDE	Stochastic differential equation

Author details

Neil K. Chada
King Abdullah University of Science and Technology, Thuwal, Saudi Arabia

*Address all correspondence to: neilchada123@gmail.com

References

[1] Kaipio J, Somersalo E. Statistical and Computational Inverse Problems. New York: Springer Verlag; 2004

[2] Stuart AM. Inverse problems: A Bayesian perspective. Acta Numer. 2010; **19**:451-559

[3] Tarantola A. Inverse Problem Theory and Methods for Model Parameter Estimation. Philadelphia: SIAM; 1987

[4] Lorenc AC. Analysis methods for numerical weather prediction. Quarterly Journal of the Royal Meteorological Society. 1986;**112**(474):1177-1194

[5] Majda A, Wang X. Non-linear Dynamics and Statistical Theories for Basic Geophysical Flows. Cambridge University Press; 2006

[6] Oliver D, Reynolds AC. Liu N. 1st ed. Cambridge University Press: Inverse Theory for Petroleum Reservoir Characterization and History Matching; 2008

[7] Bucy RS. Nonlinear filtering theory. IEEE Transactions on Automatic Control. 1965;**10**(198):198

[8] Kalman RE. A new approach to linear filtering and prediction problems. Transactions ASME (Journal of Basic Engineering). 1960;**82**:35-45

[9] Bain A, Crisan D. Fundamentals of Stochastic Filtering. New York: Springer; 2009

[10] Law KJH, Stuart AM, Zygalakis K. Data assimilation: A mathematical introduction. In: Texts in Applied Mathematics. Springer; 2015

[11] Evensen G. Data Assimilation: The Ensemble Kalman Filter. Springer; 2009

[12] Evensen G. The ensemble Kalman filter: Theoretical formulation and practical implementation. Ocean dynamics. 2003;**53**(4):343-367

[13] Iglesias MA, Law KJH, Stuart AM. Ensemble Kalman methods for inverse problems. Inverse Problems. 2013;**29**

[14] Li G, Reynolds AC. Iterative ensemble Kalman filters for data assimilation. SPE Journal. 2009;**14**:496-505

[15] Lehtinen MS, Paivarinta L, Somersalo E. Linear inverse problems for generalised random variables. Inverse Problems. 1989;**5**(4):599-612

[16] Schillings C, Stuart AM. Analysis of the ensemble Kalman filter for inverse problems. SIAM Journal on Numerical Analysis. 2017;**55**(3):1264-1290

[17] Blomker D, Schillings C, Wacker P, Weissmann S. Well posedness and convergence analysis of the ensemble Kalman inversion. Inverse Problems. 2019;**35**(8):085007

[18] Blomker D, Schillings C, Wacker P, Weissmann S. Continuous time limit of the stochastic ensemble Kalman inversion: Strong convergence analysis. Preprint arXiv:2107.14508. 2021

[19] Benning M, Burger M. Modern regularization methods for inverse problems. Acta Numer. 2018;**27**:1-111

[20] Engl HW, Hanke K, Neubauer A. Regularization of Inverse Problems, Mathematics and its Applications. Vol. 375. Dordrecht: Kluwer Academic Publishers Group; 1996

[21] Iglesias MA. A regularising iterative ensemble Kalman method for

PDE-constrained inverse problems. Inverse Problems. 2016;**32**

[22] Iglesias MA, Yang Y. Adaptive regularisation for ensemble Kalman inversion with applications to non-destructive testing and imaging. arXiv preprint arXiv:2006.14980. 2020

[23] Chada NK, Tong XT, Stuart AM. Tikhonov regularization for ensemble Kalman inversion. SIAM Journal on Numerical Analysis. 2020;**58**(2): 1263-1294

[24] Chada NK, Tong XT. Convergence acceleration of ensemble Kalman inversion in nonlinear settings. Mathematics of Computation. 2022; **91**(335):1247-1280

[25] Lee Y. l_p regularization for ensemble Kalman inversion. SIAM Journal on Scientific Computing. 2021;**43**(5): 3417-3437

[26] Schneider T, Stuart AM, Wu J-L. Imposing sparsity within ensemble Kalman inversion. arXiv preprint, arXiv: 2007.06175. 2020

[27] Weissmann S, Chada NK, Schillings C, Tong XT. Adaptive Tikhonov strategies for stochastic ensemble Kalman inversion. Inverse Problems. 2022;**38**(4):045009

[28] Ernst OG, Sprungk B, Starkloff H-J. Analysis of the ensemble and polynomial chaos Kalman filters in Bayesian inverse problems. SIAM/ASA Journal on Uncertainty Quantification. 2015;**3**: 823-851

[29] Garbuno-Inigo A, Hoffmann F, Li W, Stuart AM. Interacting Langevin diffusions: Gradient structure and ensemble Kalman sampler. SIAM Journal on Applied Dynamical Systems. 2020; **19**(1):412-441

[30] Pavliotis G. Stochastic processes and applications: Diffusion processes, the Fokker–Planck and Langevin equations. In: Texts in Applied Mathematics. New York: Springer; 2014

[31] Kullback R, Leibler S. On information and sufficiency. Annals of Mathematical Statistics. 1951;**22**:79-86

[32] Markowich PA, Villani C. On the trend to equilibrium for the Fokker Planck equation: An interplay between physics and functional analysis, Physics and Functional Analysis, Matematica Contemporanea (SBM). 19, 1999

[33] Garbuno-Inigo A, Nüsken N, Reich S. Affine invariant interacting Langevin dynamics for Bayesian inference. SIAM Journal on Applied Dynamical Systems. 2020;**19**(3): 1633-1658

[34] Reich S, Weissmann S. Fokker–Planck particle Systems for Bayesian Inference: Computational approaches. SIAM Journal of Uncertainty Quantification. 2021;**9**(2):446-482

[35] Ding Z, Li Q. Ensemble Kalman sampler: Mean-field limit and convergence analysis. SIAM Journal on Mathematical Analysis. 2021;**53**(2): 1546-1578

[36] Kovachki NB, Stuart AM. Ensemble Kalman inversion: A derivative-free technique for machine learning tasks. Inverse Problems. 2019;**35**(9):095005

[37] Haber E, Lucka F, Ruthotto L. Never look back - A modified EnKF method and its application to the training of neural networks without back propagation. arxiv preprint. 1805;**08034**: 2018

[38] Guth PA, Schillings C, Weissmann S. Ensemble Kalman filter for neural

network based one-shot inversion. arXiv preprint. 2020

[39] Tso CM, Iglesias M, Wilkinson P, Kuras O, Chambers J, Binley A. Efficient multiscale imaging of subsurface resistivity with uncertainty quantification using ensemble Kalman inversion. Geophysical Journal International. 2021;**25**(2)

[40] Muir JB, Tsai VC. Geometric and level set tomography using ensemble Kalman inversion. Geophysical Journal International. 2020;**220**:967-980

[41] Iglesias MA, Sawlan Z, Scavino TR, Wood C. Bayesian inferences of the thermal properties of a wall using temperature and heat flux measurements. International Journal of Heat and Mass Transfer. 2018;**116**: 417-431

[42] De Simon L, Iglesias MA, Jones B, Wood C. Quantifying uncertainty in thermal properties of walls by means of Bayesian inversion. Energy and Buildings. 2018;**177**(2):177

[43] Iglesias M, Park M, Tretyakov MV. Bayesian inversion in resin transfer modelling. Inverse Problems. 2018;**34**(10)

[44] Sungkono S, Apriliani E, Saifuddin N, Fajriani F, Srigutomo W. Ensemble Kalman inversion for determining model parameter of self-potential data in the mineral exploration. In: Biswas A, editor. Self-Potential Method: Theoretical Modeling and Applications in Geosciences. Cham: Springer Geophysics. Springer; 2021

[45] Dunbar ORA, Garbuno-Inigo A, Schneider T, Stuart AM. Calibration and uncertainty quantification of convective parameters in an idealized GCM. Journal of Advances in Modeling Earth Systems. 2021

[46] Huang J, Li Z, Wang B. A Bayesian level set method for the shape reconstruction of inverse scattering problems in elasticity. Computers & Mathematics with Applications. 2021; **97**(1):18-27

[47] Albers DJ, Blancquart PA, Levine ME, Seylabi EE, Stuart AM. Ensemble Kalman methods with constraints. Inverse Problems. 2019; **35**(9):095007

[48] Chada NK, Chen Y, Sanz-Alonso D. Iterative ensemble Kalman methods: A unified perspective with some new variants. Foundations of Data Science. 2021;**3**(3):331-369

[49] Chada NK, Iglesias MA, Roininen L, Stuart AM. Parameterizations for ensemble Kalman inversion. Inverse Problems. 2018;**34**

[50] Chada NK, Schillings C, Weissmann S. On the incorporation of box-constraints for ensemble Kalman inversion. Foundations of Data Science. 2019;**1**(4):433-456

[51] Tong XT, Morzfield M. Localized ensemble Kalman inversion. arXiv preprint arXiv:2201.10821. 2022

[52] Lord G, Powell CE, Shardlow T. An Introduction to Computational Stochastic PDEs. Cambridge Texts in Applied Mathematics; 2014

Chapter 3

Nanosatellites: The Next Big Chapter in Atmospheric Tomography

Gregor Moeller

Abstract

Nanosatellite technology opens up new possibilities for earth observation. In the next decade, large satellite constellations will arise with hundreds, up to thousand of satellites in low earth orbit. A number of satellites will be equipped with rather low-cost sensors, such as GNSS receivers, suited for atmospheric monitoring. However, the future evolution in atmospheric science leans not only on densified observing systems but also on new, more complex analysis methods. In this regard, tomographic principles provide a unique opportunity for sensor fusion. The difficulty in performing the conversion of integral measurements into 3D images is that the signal ray path is not a straight line and the number of radio sources and detectors is limited with respect to the size of the object of interest. Therefore, the inverse problem is either solved linearly or iterative nonlinear. In this chapter, an overview about the individual solving techniques for the tomographic problem is presented, including strategies for removing deficiencies of the ill-posed problem by using truncated singular value decomposition and the L-curve technique. Applied to dense nanosatellite formations, a new quality in the reconstruction of the 3D water vapor distribution is obtained, which has the potential for leading to further advances in atmospheric science.

Keywords: GNSS, radio occultation, nanosatellites, singular value decomposition, wet refractivity

1. Introduction

For the reconstruction of two- or three-dimensional structures from integral measurements of atmospheric excess phase, e.g. as obtained from signals of the Global Navigation Satellite Systems (GNSS), a technique called atmospheric tomography has been invented. The basic mathematics behind was introduced by Johann Radon in 1917 and is therefore also known as the Radon transform [1]. Its first realization in form of an axial scanning computer tomograph for cross-sectional imaging of the human body was awarded in 1979 with the Nobel prize for medicine [2, 3]. Around the same time, the tomography concept was utilized for applications in geosciences. One of the very first results was communicated by [4] in 1977, who describe a three-dimensional inversion method for simultaneous reconstruction of seismic body wave velocities and epicenter coordinates.

According to [5] the basic mathematical principle of tomography is described as follows:

$$f_s = \int_S g(s) \cdot ds \tag{1}$$

where f_s is the integral function, $g(s)$ is the object property function, and ds is a small element of the ray path S along which the integral is determined.

In atmospheric tomography, $g(s)$ is typically replaced by refractivity n, which is connected to signal velocity v_p by the constant speed of light c.

$$v_p = \frac{c}{n} \tag{2}$$

In vacuum $n = 1$, in matter is $n \neq 1$, i.e. the signal is slower or faster than the speed of light, dependent on the electric and magnetic properties of the medium. The integral measure f_s is usually replaced by excess phase[1] or signal travel time which can be converted to excess phase by multiplying with the speed of light.

One difficulty in performing the integral is that the signal path through the atmosphere depends on the object properties along the signal path and is, therefore, not a straight line. A change in atmospheric conditions leads to a change in S and integral function f_s. Another challenge is related to the distribution of the radio sources and the number of detectors with respect to the size of the object of interest. From single satellite missions, the distribution of integral measurements is not optimal for the reconstruction of three-dimensional structures in an analytical way using the Radon transform. To overcome this limitation, in atmospheric sounding the Abel transform [6], a further simplification of the Radon transform, is generally applied. It allows for the determination of one-dimensional profiles of refractivity from measurements of excess phase, assuming spherical symmetry. In 1965, this technique was applied to measurements of the Mariner four spacecraft to study important properties of the Martian atmosphere and is nowadays commonly applied to GNSS phase measurements obtained from dedicated radio occultation missions [7–9]. However, standard processing strategies based on the Abel transform do not allow for resolving horizontal features in the atmosphere. With the advent of nanosatellite technology, the number and distribution of signals have significantly increased - leading to the situation that the assumptions made to derive the Abel transform (spherical symmetry and parallel observation paths) become a limiting factor in the analysis of space-based radio occultation observations. To overcome this limitation, the existing observations can be stacked together to solve Eq. (1) either linearly or iterative non-linearly [5]. A complete non-linear solution is difficult to achieve but also not necessary for most applications in geoscience since it can be demonstrated that the signal path is not significantly perturbed by linearization assumptions. In Section 2, common solving techniques (linear and non-linear) are presented, and in Section 3 and Section 4, it is analyzed whether they can be utilized to reconstruct refractivity fields in the neutral atmosphere from GNSS measurements of atmospheric excess phase on-board dense nanosatellite formations.

---

[1] In literature this quantity has been given many different names, such as *atmospheric excess phase*, *atmospheric phase delay* or derivations thereof.

2. Solving techniques

If f_s in Eq. (1) is replaced by the atmospheric excess phase (aep) and $g(s)$ by the relation $n - 1$, the basic function of atmospheric tomography is obtained as follows:

$$aep = \int_S n \cdot ds - \int_{S_0} ds \tag{3}$$

where S is the "true" signal path and S_0 is the theoretical straight line signal path in a vacuum. The second term on the right-hand side of Eq. (3) stems from the definition of aep, describing only the atmospheric contribution to the excess phase. Strictly speaking also the limits of the integral have to be adapted to the relevant parts in the atmosphere, e.g. up to about $80km$ altitude for the neutral atmosphere.

Fermat's principle tells us that first-order changes in the signal path cause second-order changes in signal travel time, i.e. for small variations in the object properties, the travel time is stationary. This principle is very beneficial since it allows to define two simplified versions of atmospheric tomography, the so-called linear and non-linear approach. In linear tomography, the bent signal path S is replaced by a straight line S_0 and corrections to n are made by ignoring atmospheric bending. In contrast, the iterative non-linear approach takes the signal bending into account by the definition of the ray paths but not in the inversion of n along ds. This means after each processing step the signal paths are re-computed, e.g. by solving the so-called Eikonal using ray-tracing shooting techniques [10].

A numerical solution for Eq. (1) is obtained by discretizing the object of interest, e.g. the neutral atmosphere in area elements (in two-dimensions) or volume elements (in three-dimensions). Further, it is assumed that in each volume element the index of refraction is constant.[2] In the atmosphere, the index of refraction n is close to 1, thus it can be replaced by refractivity $N = (n - 1) \cdot 10^6$. With these adaptions Eq. (1) reads:

$$aep = \sum_{k=1}^{m} N_k \cdot d_k \tag{4}$$

where N_k is the constant refractivity and d_k is the ray length in the volume element k. Assuming l observations, indexed by $j = 1,2, \ldots ,l$ and m volume elements (short: voxels), indexed by $k = 1,2, \ldots ,m$, the individual observation equations can be combined into a linear equation system. In matrix notation the resulting tomographic equation reads:

$$\boldsymbol{AEP} = \boldsymbol{A} \cdot \boldsymbol{N} \tag{5}$$

where $\boldsymbol{AEP}$ is the observation vector of size $(l, 1)$ and $\boldsymbol{N}$ is the unknown vector of size $(m, 1)$ describing the properties in each volume element k. The (l, m) matrix $\boldsymbol{A}$ contains the spatial derivatives of the observations aep_j with respect to the unknowns N_k.

[2] In recent works by [11] or [12] alternative parameterizations, such as a trilinear, spline or adaptive node parameterizations are suggested for a more accurate description of the refractivity distribution without considerably increasing the number of unknowns.

$$A = \begin{bmatrix} \frac{\partial aep_1}{\partial N_1} & \frac{\partial aep_1}{\partial N_2} & \cdots & \frac{\partial aep_1}{\partial N_m} \\ \frac{\partial aep_2}{\partial N_1} & \frac{\partial aep_2}{\partial N_2} & \cdots & \frac{\partial aep_2}{\partial N_m} \\ \vdots & \vdots & \vdots & \vdots \\ \frac{\partial aep_l}{\partial N_1} & \frac{\partial aep_l}{\partial N_2} & \cdots & \frac{\partial aep_l}{\partial N_m} \end{bmatrix} \tag{6}$$

Since Eq. (4) is linear, the partial derivatives of *aep* are the ray lengths (d_k) in each voxel. For linear tomography, d_k is computed from the line of sight vector between the transmitter and receiver. In the non-linear approach, the bending of the signal is taken into account, e.g. by using ray tracing shooting techniques [13]. Therefore, a priori information about the atmospheric state (e.g. in the form of numerical weather forecasts) is needed. Dependent on the quality of the a priori model, additional iterations might be necessary. After each iteration, the estimated refractivity field is considered as input for the ray tracer. The processing is repeated until the determination of the ray path converges. This happens usually after 1–2 iterations.

2.1 The inverse problem

An analytical solution for the tomographic equation (Eq. (5)) can be found by the inversion of matrix A.

$$N = A^{-1} \cdot AEP \tag{7}$$

The inverse A^{-1} exists if the determinant of A is non-zero. This requires that A is a squared matrix ($l = m$). Otherwise, the matrix A is called singular, i.e. does not have a matrix inverse. Unfortunately latter is the case in most atmospheric tomography applications since the observation data, e.g. the GNSS measurements, are considered as “incomplete”. Therefore, the matrix A has zero singular values and Eq. (7) becomes ill-posed. In literature, several strategies are described, which allow to remove the deficiencies of the ill-posed problem. They either try to solve the inverse problem or to avoid it. The most prominent ones are:

- Iterative algebraic reconstruction techniques
- Truncated singular value decomposition
- Tikhonov regularization

These three techniques have been selected since they were proven in practice as the most reliable, as described briefly in the following subsections.

2.1.1 Algebraic reconstruction techniques

The iterative algebraic reconstruction technique (ART) has been suggested in 1937 by [14] for solving linear equation systems. This technique avoids the inversion problem and initializes the matrix A row-wise. This is very beneficial for large equation systems. Applied to Eq. (7) the ART algorithm reads:

$$N^{i+1} = N^i + \frac{\omega}{\langle A_j, A_j \rangle} \cdot \left(aep_j - \left\langle A_j^T,\ N^i \right\rangle \right) \cdot A_j \tag{8}$$

where A_j indicates the jth row of the matrix A, $\langle A_j, A_j \rangle$ is the resulting inner product and the difference $aep_j - \left\langle A_j^T, N^i \right\rangle$ is the prefit-residual from the last iteration. Based on the number of traversed volume elements and the relaxation factor ω the residual is split into multiple components and applied to N^i in order to obtain the improved refractivity field N^{i+1}, which is again input for the next iteration. The processing is stopped once Eq. (8) converges to the solution of Eq. (7) with minimal norm if $0 < \omega < 2$. For ground-based GNSS networks, the best results have been obtained with a relaxation parameter of ~ 0.175, see [15]. Studies from [16] or [17] manifest that the algorithms of the ART family are also very successful in reconstructing the total electron content (TEC) in the ionosphere. Dependent on how the discretization is done, different realizations of ART exist. For tomographic reconstruction especially the multiplicative algebraic reconstruction techniques (MART) and the simultaneous iterative reconstruction techniques (SIRT) are worth mentioning, see [18] or [19]. In contrast to the original ART algorithm, MART leads in general to faster convergence and SIRT has the benefit of being impervious to the order of measurements (aep_j).

2.1.2 Truncated singular value decomposition

For ill-conditioned least squares problems, [20, 21] invented a general solution, widely known as pseudo inverse or Moore-Penrose inverse $\boldsymbol{A}^+$. A numerical solution for the pseudo inverse can be obtained by singular value decomposition [22]. This requires a split of the matrix $\boldsymbol{A}$ into three components as follows:

$$\boldsymbol{A} = \boldsymbol{U} \cdot \boldsymbol{S} \cdot \boldsymbol{V}^T \tag{9}$$

with $\boldsymbol{U}$ (l,l) and $\boldsymbol{V}^T\,(m,m)$ as orthogonal matrices, containing the normalized left and right singular vectors of $\boldsymbol{A}$, respectively. Matrix $\boldsymbol{S}$ (l,m) is a diagonal matrix with singular values $s_{k,k}$ arranged in descending order. By using only the non-zero diagonal elements of $\boldsymbol{S}$ the pseudo inverse is obtained as follows:

$$\boldsymbol{A}^+ = \boldsymbol{V} \cdot \boldsymbol{S}^{-1} \cdot \boldsymbol{U}^T \tag{10}$$

The 2-norm of the matrix $\mathbf{S}$ defines the condition number $\kappa(\mathbf{A})$. It can be interpreted as the ratio between the largest and the smallest singular values.

$$\kappa(\mathbf{A}) = \frac{|s_{max}|}{|s_{min}|} \tag{11}$$

A well-conditioned matrix has a condition number $\kappa(\boldsymbol{A})$ near 1. The resulting tomography solution is rather insensitive to measurement errors. A large condition number indicates an ill-conditioned problem. According to Eq. (11), the condition number of $\boldsymbol{A}$ improves by neglecting tiny singular values. This technique is known as truncated singular value decomposition (TSVD), see [23]. It allows to approximate the ill-conditioned matrix $\boldsymbol{A}$ by a matrix $\tilde{\boldsymbol{A}}$ of lower rank.

For atmospheric tomography, [24] suggested $s_{lim} = 2.8km$ as the threshold for $s_{k,k}$, i.e. all singular values smaller than s_{lim} are set to zero. However, in practice an optimal threshold for s_{lim} can be determined using the L-curve technique [23]. Therefore, a set of solutions is determined with varying s_{lim}-values. For each solution, the 2-norm of the estimated parameters is plotted against the 2-norm of the residuals. By connecting all points in a log–log plot a concave L-shaped curve is obtained, whereby the corner of the curve, i.e. the point of maximum curvature, defines the optimal threshold (s_{opt}) for singular value decomposition. **Figure 1** shows an L-curve for a typical GNSS tomography setting. In this example, the optimal solution is obtained by setting s_{opt} to 0.032. This point was found by testing various values for s_{lim} between $3 \cdot 10^{-1}$ and $3 \cdot 10^{-5}$. After each processing step, the solution norm $log \|\boldsymbol{N}\|_2$ is plotted against the residual norm $log \|\boldsymbol{A} \cdot \boldsymbol{N} - \boldsymbol{AEP}\|_2$ and the corner point of the resulting curve is marked as the optimal solution.

2.1.3 Tikhonov regularization

A more generalized solution to the regularization problem can be found in [25], who describes the minimization problem as follows:

$$\boldsymbol{N}_\eta = arg\ min \left\{ \|\boldsymbol{A} \cdot \boldsymbol{N} - \boldsymbol{AEP}\|_2^2 + \eta^2 \|\boldsymbol{L}(\boldsymbol{N}_0 - \boldsymbol{N})\|_2^2 \right\} \tag{12}$$

where η is called the regularization parameter or Tikhonov factor and $\boldsymbol{N}_0$ is an approximation of $\boldsymbol{N}$. The "size" of the solution is defined by the norm $\|\boldsymbol{L}(\boldsymbol{N}_0 - \boldsymbol{N})\|_2$ and the "fit" by the norm of the residual vector $\|\boldsymbol{A} \cdot \boldsymbol{N} - \boldsymbol{AEP}\|_2$. One possibility to solve Eq. 12 is to treat it as a least squares problem. In [26] it is shown that the matrix $\boldsymbol{L}$ can be replaced by an identity matrix $\boldsymbol{I}$, i.e. the condition number of $\boldsymbol{A}$ is improved by adding a small multiple of the identity to the matrix $\boldsymbol{A}$.

$$\tilde{\boldsymbol{A}} = \boldsymbol{A} + \eta \cdot \boldsymbol{I} \tag{13}$$

A possible solution for η can be obtained by means of singular value decomposition (see Section 2.1.2). Thereby the elements of the diagonal matrix $\boldsymbol{S}$ are replaced by the coefficients $r_{k,k}$.

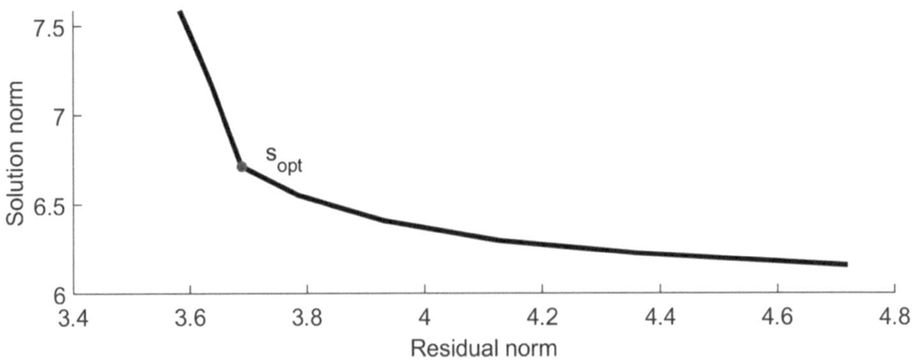

Figure 1.
Representative L-curve for a typical GNSS tomography inversion problem. The red dot indicates the corner point of the L-curve and therewith the optimal tomography solution.

$$r_{k,k} = \frac{s_{k,k}^2}{s_{k,k}^2 + \eta} \tag{14}$$

If the Tikhonov factor is defined as a sharp filter

$$\eta = \begin{cases} 1 for s_{k,k} \geq s_{lim} \\ 0 for s_{k,k} < s_{lim} \end{cases} \tag{15}$$

the resulting solution can be interpreted as smoothed TSVD solution.

2.2 The partial least squares solution

By treating Eq. (7) as a least squares problem, the basic equation of the weighted least squares estimator reads:

$$\hat{\boldsymbol{N}} = \left(\boldsymbol{A}^T \cdot \boldsymbol{P} \cdot \boldsymbol{A}\right)^{-1} \cdot \boldsymbol{A}^T \cdot \boldsymbol{P} \cdot \boldsymbol{AEP} \tag{16}$$

where $\boldsymbol{P}$ is a weighting matrix, which allows to take the relative accuracy and possible constraints between the observations into account. The least squares solution $\hat{\boldsymbol{N}}$ is obtained by minimizing the 2-norm of the observation residuals. Thereby we assume, that the observations are normally distributed, i.e. free of gross errors or systematic effects.

By combining Eq. (10) with Eq. (16) the tomography solution reads:

$$\hat{\boldsymbol{N}} = \boldsymbol{V} \cdot \boldsymbol{S}^{-1} \cdot \boldsymbol{U}^T \cdot \boldsymbol{A}^T \cdot \boldsymbol{P} \cdot \boldsymbol{AEP} \tag{17}$$

where the columns of $\boldsymbol{U}$ and $\boldsymbol{V}^T$ are the normalized left and right singular vectors of $\boldsymbol{A}^T \cdot \boldsymbol{P} \cdot \boldsymbol{A}$, respectively and matrix $\boldsymbol{S}$ (l,m) is the diagonal singular value matrix as defined in Section 2.1.2.

2.2.1 The *a priori* field

In Section 2, the linear and non-linear approaches have been defined to reconstruct the GNSS signal paths through the atmosphere. While the linear approach is not dependent on any external data, the non-linear approach requires an a priori refractivity field, e.g. derived from the standard atmosphere or numerical weather forecasts, to reconstruct the bent signal path. Besides, the a priori field can be also utilized to stabilize the tomographic equation system. One possibility is to treat the additional information ($\boldsymbol{N}_0$) as absolute constraints:

$$\hat{\boldsymbol{N}} = \boldsymbol{N}_0 + \boldsymbol{V} \cdot \boldsymbol{S}^{-1} \cdot \boldsymbol{U}^T \cdot \boldsymbol{A}^T \cdot \boldsymbol{P} \cdot (\boldsymbol{AEP} - \boldsymbol{A} \cdot \boldsymbol{N}_0) \tag{18}$$

In the following, this solution is called the **constrained solution.**

Another possibility to handle the extended equation system is to treat it as a system of subsets with

$$\boldsymbol{A}_{ext} = \begin{bmatrix} \boldsymbol{A} \\ \boldsymbol{A}_c \end{bmatrix} \tag{19}$$

$$AEP_{ext} = \begin{bmatrix} AEP \\ N_0 \end{bmatrix} \quad (20)$$

$$P_{ext} = \begin{bmatrix} P \\ P_c \end{bmatrix} \quad (21)$$

where A_c is the design matrix and P_c the weighting matrix for N_0. The extended equation system can be solved using Eq. (17), whereby A, AEP and P are replaced by its extended complements A_{ext}, AEP_{ext} and P_{ext}. In principle, it provides the same results as the constrained solution.

A third possibility would be to solve Eq. (18) separately for each observation type using the estimates ($\hat{N}$) and the variance–covariance matrix of the estimates ($Cov_{\hat{N}N}$) from the first step as a priori information for the next step. In the case of two subsets the corresponding tomography solution reads:

$$\hat{N}_1 = V \cdot S^{-1} \cdot U^T \cdot A^T \cdot P \cdot AEP \quad (22)$$

$$Cov_{\hat{N}N} = V \cdot S^{-1} \cdot U^T \quad (23)$$

where U, V and S are obtained by singular value decomposition of the matrix $A^T \cdot P \cdot A$. For the second (final) solution both, $\hat{N}_1$ and $Cov_{\hat{N}N}$ are introduced into the equation system as follows:

$$\hat{N} = \hat{N}_1 + V \cdot S^{-1} \cdot U^T \cdot A_c^T \cdot P_c \cdot \left(N_0 - A_0 \cdot \hat{N}_1\right) \quad (24)$$

with S, U and V obtained by truncated singular value decomposition (see subsection 2.1.2) of the matrix $A_0^T \cdot P_0 \cdot A_0 + Cov_{\hat{X}X}^{-1}$. In the following this solution is called the **partial solution**. In case the matrix A is of full rank or if only one set of observations is available, the constrained solution and the partial solution provide identical results. In the case of an ill-conditioned matrix, the partial solution has the advantage that the eigenvalue can be computed for each subset of observations. In large equation systems, this allows to reduce computational load since the matrix A is divided into several parts.

2.2.2 Observation weights

Up to now, the individual observations were considered as uncorrelated and equally accurate. However, for varying input data it might be beneficial to set up a weighting matrix. In case the relative accuracy between observations is known, they can be directly introduced into the equation system by defining the weighting matrix

$$P = \sigma_0^2 \cdot Cov_{ll}^{-1} \quad (25)$$

where variance co-variance matrix Cov_{ll} reflects the precision of the observations on its diagonal elements (σ_n^2) with σ_0^2 as the a priori variance of the unit weight. In the case of uncorrelated observations and unit variances, the matrix Cov_{ll} simplifies to an identity matrix of size (n, n).

In case no accurate information is available, a weighting model can be utilized. For ground-based GNSS observations, an elevation-dependent weighting is common, for satellite-to-satellite observations a weighting based on carrier-to-noise density $C/N0$

$\frac{\partial N}{\partial T}\Delta T$	$\frac{\partial N}{\partial p}\Delta p$	$\frac{\partial N}{\partial e}\Delta e$	ΔN
$\pm 0.25ppm$	$\pm 0.08ppm$	$\pm 2.37ppm$	$\pm 2.39ppm$

Table 1.
Standard deviation of refractivity and its components, assuming a typical meteo sensor error of 0.3hPa for pressure, 0.2K for temperature and 3% for relative humidity - computed for standard atmospheric conditions at sea level ($p = 1013hPa, T = 15°C, rh = 60\%$).

seems to be more useful since the observations are usually gathered around 0deg the elevation angle or below. For the a priori refractivity field, a height-dependent weighting model is recommended. By focusing on the neutral atmosphere and assuming that Eq. (26) is exact

$$N = K_1 \cdot \frac{p_d}{T} + K_2 \cdot \frac{e}{T} + K_3 \cdot \frac{e}{T^2} \tag{26}$$

the theoretical standard deviation for refractivity reads[3]:

$$\sigma_N = \left[\left(\frac{\partial N}{\partial T} \cdot \sigma_T \right)^2 + \left(\frac{\partial N}{\partial p} \cdot \sigma_p \right)^2 + \left(\frac{\partial N}{\partial e} \cdot \sigma_e \right)^2 + \ldots \right]^{\frac{1}{2}} \tag{27}$$

For in-situ measurements, [27] provides theoretical standard deviations for pressure p, temperature T, and water vapor pressure[4] e for a wide range of meteorological sensors. In addition, height-dependent error curves for the three meteorological parameters can be obtained from [28]. The resulting uncertainties, assuming standard atmospheric conditions at sea level, are listed in **Table 1**.

The standard deviation of refractivity is $2.39ppm$, which equates to a relative uncertainty of 0.75%. By far the largest impact ($2.37ppm$) is related to the uncertainty of water vapor. Consequently, the utmost care has to be taken when measuring humidity and temperature.

3. Observations of atmospheric excess phase

Satellite refractometric sounding of the atmosphere and the underlying inverse problems have been under investigation since the 1960s, see [29–31]. However, it was not until 1976 that the first radio occultation (RO) experiment was carried out to survey the earth's atmosphere within the Apollo-Soyuz mission [32]. Until then, the major problem noted was the lack in accuracy of refractometric measurements of phase or Doppler shift [33]. This limitation has widely been overcome with the emergence of the Global Positioning System (GPS) around the 1980s [7]. Since the proof of concept during the GPS-MET satellite mission in 1995 various satellites have been equipped with precise GNSS radio occultation receivers, leading to approximately $500 - 600$ globally distributed radio occultation profiles per day and satellite, assuming a 32-satellite GPS constellation.

---

3 Assuming that the uncertainty of the refractivity constants is negligible.

4 Since water vapor pressure is usually not measured directly, it can be computed from relative humidity and temperature.

3.1 The observation equation

The phase that a receiver obtains from a GNSS satellite can be modeled as

$$L^s_{r,\nu} = \varrho^s_r + c \cdot \delta t_r - c \cdot \delta t^s + \Delta\varrho^s_{r,\nu} + \nu \cdot n^s_{r,\nu} \quad (28)$$

with

$L^s_{r,\nu}$	... phase observation for the transmitter-receiver pair (s-r) at frequency ν
ϱ^s_r	... geometric distance between transmitter s and receiver r
$c \cdot \delta t_r$	... correction of receiver clock at signal reception time t
$c \cdot \delta t^s$	... correction of satellite clock at transmission time $t - \tau^s_r$
$\Delta\varrho^s_{r,\nu}$	... all delays due to propagation effects
$n^s_{r,\nu}$	... unknown integer number of cycles (carrier phase ambiguity)

The objective of refractometric sounding is to extract the atmospheric propagation effects from the phase observations. Assuming that relativistic effects, satellite-specific multipath effects and antenna-specific phase center corrections are known and removed, the remaining effects in $\Delta\varrho^s_{r,\nu}$ can be divided into two terms:

$$\Delta\varrho^s_{r,\nu} = \Delta\varrho^s_{r,trp} + K \frac{TEC^s_{r,\nu}}{f^2_r} \quad (29)$$

where the first term ($\Delta\varrho^s_{r,trp}$) describes the delay of the carrier phase in the neutral atmosphere and the second term $\left(K \frac{TEC^s_{r,\nu}}{f^2_r}\right)$ the advancement of the carrier phase in the ionosphere. The integral term $TEC^s_{r,\nu}$ is the electron density along the ray path between transmitter s and receiver r, scaled by a constant term K.

A detailed description of the individual systematic effects can be found in [34]. In the following, special attention is given to the modeling and estimation of neutral atmospheric effects ($\Delta\varrho^s_{r,trp}$) assuming that the first-order ionospheric effect (up to 99.9%) can be removed by forming an ionospheric-free linear combination L_{IF}. Condition therefore is, that the receiver tracks the GNSS carrier phase simultaneous on two frequencies

$$L_{IF} = \frac{f^2_1 \cdot L^s_{r,1} - f^2_2 \cdot L^s_{r,2}}{f^2_1 - f^2_1} \quad (30)$$

where the nominal frequencies f_1 and f_2 are defined by the satellite system frequency plan (e.g. 1575.42MHz for GPS $L1$ and 1227.60MHz for GPS $L2$).

3.2 Calibration of the phase signal

For the extraction of atmospheric phase excess from phase observations, first, the phase signal has to be calibrated. Therefore, the clock effects in Eq. (28) are eliminated. This can be achieved if the occulting receiver satellite can simultaneously see an occulting GNSS and a non-occulting GNSS satellite. Elimination of the clock effects is also possible if the same GNSS satellites are visible from a ground-based GNSS receiver. In addition, one needs to precisely know the position and velocity of the transmitting and receiving satellites. The orbits for the GNSS satellites can be obtained from services such as the International GNSS service (see www.igs.org). The position and velocity of the receiving satellite have to be computed using the phase measurements recorded from the GNSS radio occultation receiver or from a dedicated POD (precise orbit determination) receiver on board the receiving satellite. For further details about the procedure of POD, the reader is referred to [35]. A more detailed description of the calibration of the phase measurements is given in, e.g. [36]. Once the signal is calibrated, the atmospheric phase excess can be used to set up the observation equations for the tomographic processing. The precondition for a stable tomography solution is, that enough overlapping observations are available. Therefore, in the following the concept of a dense nanosatellite formation is introduced.

4. Concept of a dense nanosatellite formation

4.1 Introduction

In recent years, nanosatellite technology has become increasingly important for a wide variety of applications, such as communication, technology demonstration, heliophysics, astrophysics, earth science, or planetary science [37–39]. Most of the existing mission concepts are based on the CubeSat form factor established by Professor Bob Twiggs at the Department of Aeronautics and Astronautics at Stanford University in late 1999. Although small satellites have existed since the very beginning of spaceflight, the definition of the CubeSat standard "made it possible to bring production to a level of flexibility and innovation never seen before" [40]. As of September 2022, over 2000 nanosatellites were launched into orbit, with a record number of 143 satellites launched on a single rocket on board the Transporter-1 mission in January 2021. In the next decade, we expect that the number of nanosatellite launches per year will continue to rise by a factor of 4–5, leading to dense observation networks in low earth orbit. Innovations in satellite technology, such as miniaturized GNSS receivers [41] and intelligent processing strategies will further boost the realization of new observation concepts based on nanosatellite technology and the establishment of dense satellite formations in highly interesting but yet scarcely-explored regions in the earth‘s atmosphere and beyond.

4.2 The observation geometry

The multi-frequency signals from over a hundred active GNSS satellites gathered on board each nanosatellite allow for measuring the atmospheric state with unprecedented spatiotemporal resolution. For the proof of concept, we assume a formation of four nanosatellites, injected into a polar orbit. The advantage from such a configuration is that we can get simultaneous radio occultation observations that are closely located [42]. **Figure 2** shows the observation geometry together with the ray paths through the lower $8km$ of the atmosphere.

The spacing between the nanosatellites is set to $dM = 1.9$deg (approx. $230km$). At an altitude of $550km$ this corresponds to a temporal spacing of about $30s$. Due to the limb-sounding geometry and high inclination of most nanosatellites, a global distribution can be obtained with 500 – 600 radio occultation events per nanosatellite and day assuming a 32-satellite GPS constellation.

The angle under which the GNSS satellites are observed is constantly changing. In order to characterize the observation geometry, we distinguish between two scenarios. In the first scenario, the observation angle is close to 90deg, i.e. the RO measurements are obtained in a cross-track direction. In consequence, from the four nanosatellites, we obtain ray paths that are widely parallel to each other. In the second scenario, the angle is close to 0deg or 180deg. This leads to RO measurements in the flight direction or anti-flight direction. In both cases, a unique observation geometry is obtained, in which consecutive observations overlap, as shown in **Figure 2**.

4.3 Tomography case study

At the time of writing, real GNSS measurements from a dense nanosatellite formation were not available. Thus, for technique demonstration, a closed-loop simulation was carried out using the Weather Research and Forecasting (WRF) model to simulate the atmospheric state along the GNSS radio occultation signals shown in **Figure 2**.

In the first step, the signal paths through the atmosphere were reconstructed every 500*ms* using ray-tracing shooting techniques [13] with a step size of 1*km*. For each ray point, wet refractivity (N_w) was computed from WRF temperature (T) and water vapor pressure (e) fields using Eq. (31)

$$N_w = K_2' \cdot \frac{e}{T} + K_3 \cdot \frac{e}{T^2} \tag{31}$$

where the constant K_2' is given by

$$K_2' = K_2 - K_1 \cdot \frac{M_w}{M_d} \tag{32}$$

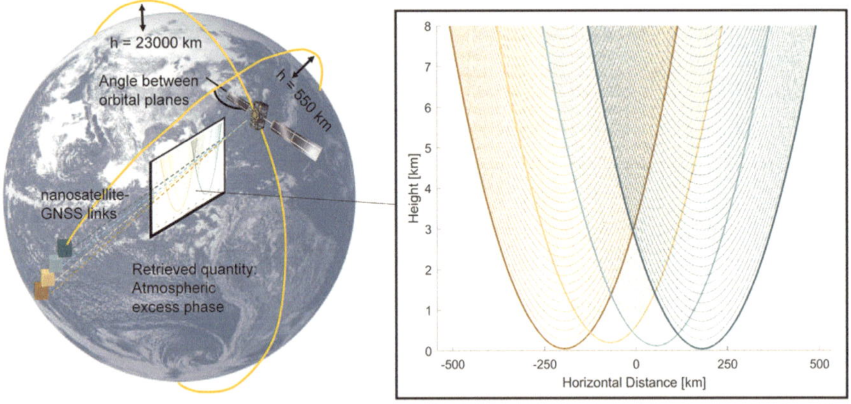

Figure 2.
Left: The observation geometry for one GNSS satellite simultaneous observed by four nanosatellites in a string-of-pearls formation. Right: The resulting radio occultation signal paths.

with $K_1 = 77.689 \frac{K}{hPa}$, $K_2 = 71.2952 \frac{K}{hPa}$ and $K_3 = 375463 \frac{K^2}{hPa}$ as the refractivity constants and M_w and M_d as the molar mass of dry air and water vapor, respectively.

Figure 3 (left) shows the resulting wet refractivity distribution in the lowest $8km$ of the atmosphere, with a water vapor inversion layer at a height of approximately $2 - 4km$. By integration along the signal paths, the wet refractivity can be converted into atmospheric excess phase using Eq. (4). **Figure 3** (right) shows that the inversion layer in the WRF model also propagates into the simulated observations of atmospheric phase excess.

To reconstruct the 2D refractivity fields from the atmospheric excess phase, the area covered by the observations was discretized in area elements with a grid size of 22 km (horizontally) and 0.2 km (vertically). The tomographic processing itself was carried out with the ATom software package [13]. **Table 2** summarizes the major settings.

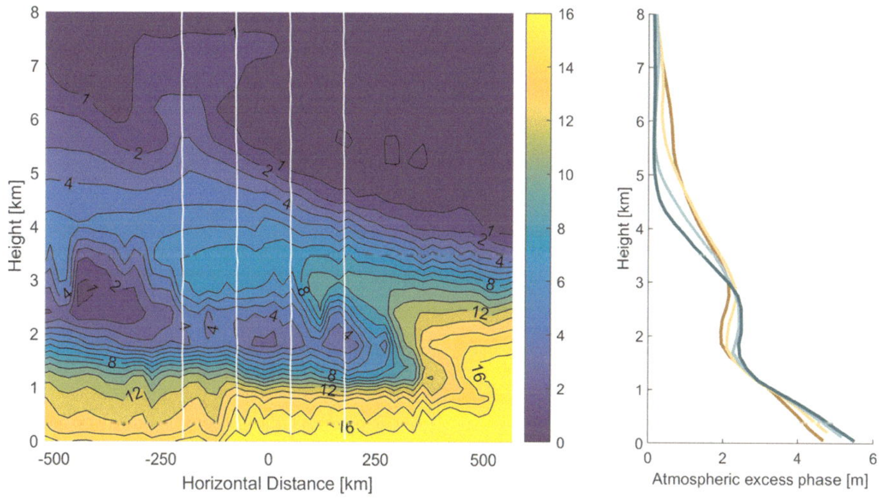

Figure 3.
Left: Weather Research and Forecasting (WRF) model derived wet refractivity fields [ppm] with the overlaying white lines showing the tangent points of the four RO ray paths through the lower 8km of the atmosphere. Right: The resulting atmospheric excess phase observations [m] by integration through the wet refractivity field.

Parameter	Settings
Case study domain	Equatorial pacific ocean ($140 - 150 \text{deg}E$)
Case study period	Late autumn 2006
Model resolution	$22km$ (horizontally) $\times$ $0.2km$ (vertically)
Tomography software	Modified version of ATom software package*
Initial field	smooth WRF field
Inversion method	Singular value decomposition ($eigenv_{min} = 0.01km^2$)
Estimation method	Iterative weighted least squares adjustment
Convergence criteria	RMS of weighted residuals

**https://github.com/GregorMoeller/ATom.*

Table 2.
Tomography settings applied for the reconstruction of refractivity fields from (simulated) RO observations.

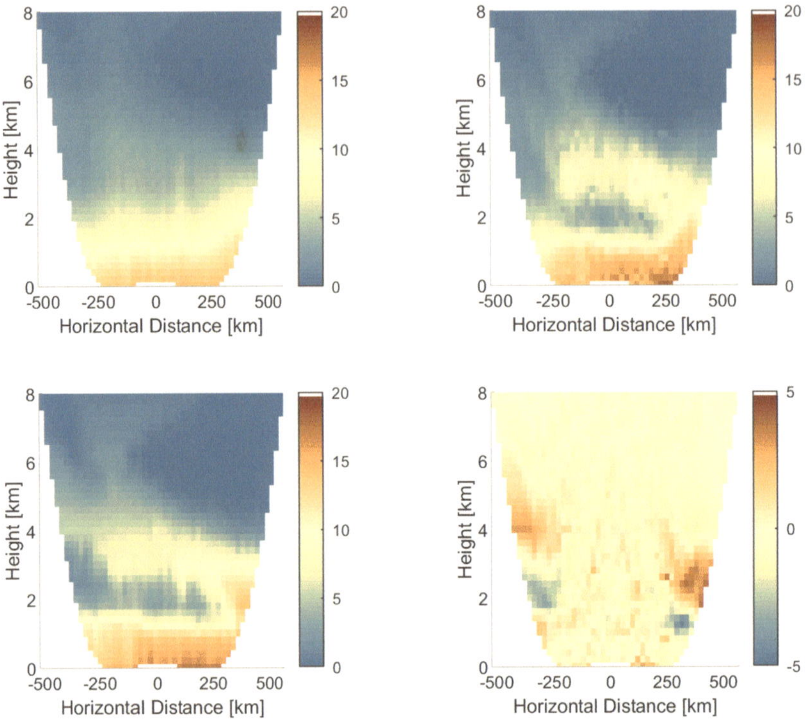

Figure 4.
Top left: Smooth WRF refractivity field used to initialize the tomography solution. Top right: Estimated refractivity field (tomography solution). Bottom left: WRF refractivity field (reference). Bottom right: Closed-loop validation (tomography minus WRF) to assess the performance of the tomography approach.

The resulting refractivity fields are visualized in **Figure 4**. The upper left plot shows the a priori field, a smooth WRF refractivity field, used to initialize the tomography solution. For the computation of the smoothed field, a sliding window filter was applied to the WRF data to remove the inversion layer and therefore, reduce the information contained in the initial field. In the upper right plot, the actual tomography solution is shown. By comparison with the WRF reference field (lower left plot), the reconstruction capabilities of the tomography approach can be assessed. The differences between the two models are shown in the lower right plot. Overall voxels, a Root Mean Square Error (RMSE) of $0.9ppm$ (21.8%) and a bias of $0.03ppm$ was received.

Overall, the best solution is obtained within the horizontal range (-250 km, 250 km) in which multiple observations overlap and therefore, help to stabilize the tomography resolution. In this core domain of the tomography model, an RMSE of $0.5ppm$ (9.6%) was obtained, which is by a factor of two better than in the outer regions.

5. Conclusions and outlook

In this chapter, the basic aspects of the remote sensing of the lower earth's atmosphere using tomography radio occultation methods are addressed. My motivation was to provide an overview about the current achievements in tomographic processing and its potential for the processing of radio occultation measurements collected from very light-weight and power-efficient GNSS sensors onboard dense nanosatellite formations. In a number of closed-loop validations, the expected observations have been analyzed and possible processing strategies have been evaluated. Due to the unique observation geometry, combined processing of overlapping radio occultation measurements using tomographic principles is possible and allows to generate high-resolution cross-sections of the lower atmosphere. Thus, I believe that tomography products have great potential to advance current knowledge, e.g. as a weather analysis tool or as a complementary observation technique for water vapor distribution, which can be assimilated into operational weather forecast systems. Once the required sensor technology is available, not only the communication industry but also the earth observation community will benefit from new observation concepts based on nanosatellite technology. If the proposed observation concept is also suited for the monitoring of the ionosphere has to be evaluated in future studies.

Conflict of interest

The authors declare no conflict of interest.

Abbreviations

AEP	Atmospheric Excess Phase
ART	Algebraic Reconstruction Technique
ATom	Atmospheric Tomography software package
GNSS	Global Navigation Satellite Systems
MART	Multiplicative Algebraic Reconstruction Technique
POD	Precise Orbit Determination
RMSE	Root Mean Square Error
RO	Radio Occultation
SIRT	Simultaneous Iterative Reconstruction Technique
TEC	Total Electron Content
TSVD	Truncated Singular Value Decomposition
WRF	Weather Research and Forecasting model

Author details

Gregor Moeller
ETH Zurich, Institute of Geodesy and Photogrammetry, Zurich, Switzerland

*Address all correspondence to: gmoeller@ethz.ch

References

[1] Radon J. Über die Bestimmung von Funktionen durch Ihre Integralwerte längs gewisser Mannigfaltigkeiten. Berichte über die Verhandlungen der Sächsischen Gesellschaft der Wissenschaften zu Leipzig. 1917;**69**:262-277

[2] Cormack AM. Representation of a function by its line integrals, with some radiological applications. Journal of Applied Physics. 1963;**34**(9):2722-2727. DOI: 10.1063/1.1729798

[3] Hounsfield GN. Computerized transverse axial scanning tomography: Part I. Description of the system. The British Journal of Radiology. 1973; **46**(552):1016-1022. DOI: 10.1259/0007-1285-46-552-1016

[4] Aki K, Christoffersson A, Husebye ES. Determination of the three-dimensional seismic structure of the lithosphere. Journal of Geophysical Research. 1977;**82**(2):277-296. DOI: 10.1029/JB082i002p00277

[5] Iyer HM, Hirahara K. Seismic Tomography: Theory and Practice. 1st ed. Dordrecht, Netherlands: Springer; 1993. p. 864. ISBN: 978-0412371905

[6] Abel NH. Auflösung einer mechanischen Aufgabe. Journal für die reine und angewandte Mathematik. 1826; **1**:153-157. DOI: 10.1515/crll.1826.1.153

[7] Ware R, Rocken C, Solheim F, Exner M, Schreiner W, Anthes R, et al. GPS sounding of the atmosphere from low earth orbit: Preliminary results. Bulletin of the American Meteorological Society. 1996;**77**:19-40. DOI: 10.1175/1520-0477(1996)077<0019:GSOTAF>2.0.CO;2

[8] Schmidt T, Wickert J, Marquardt C, Beyerle G, Reigber C, Galas R, et al. GPS radio occultation with CHAMP: An innovative remote sensing method of the atmosphere. Advances in Space Research. 2004;**33**(7):1036-1040. DOI: 10.1016/S0273-1177(03)00591-X

[9] Schreiner WS, Weiss JP, Anthes RA, Braun J, Chu V, Fong J, et al. COSMIC-2 radio occultation constellation: First results. Geophysical Research Letters. 2020;**47**:7. DOI: 10.1029/2019GL086841

[10] Moeller G, Landskron D. Atmospheric bending effects in GNSS tomography. Atmospheric Measurement Techniques. 2019;**12**:23-34. DOI: 10.5194/amt-12-23-2019

[11] Perler D. Water vapor tomography using global navigation satellite systems [doctoral thesis]. ETH Zurich: Institute of Geodesy and Photogrammetry. 2011

[12] Ding N, Zhang SB, Wu SQ, Wang X. Adaptive node parameterization for dynamic determination of boundaries and nodes of GNSS tomographic models. Journal of Geophysical Research Atmospheres. 2018;**123**(4):1990-2003. DOI: 10.1002/2017JD027748

[13] Moeller G. Reconstruction of 3D wet refractivity fields in the lower atmosphere along bended GNSS signal paths [doctoral thesis]. TU Wien: Department of Geodesy and Photogrammetry. 2017. DOI: 10.34726/hss.2017.21443

[14] Kaczmarz S. Angenäherte Auflösung von Systemen linearer Gleichungen. Bulletin International de l' Académie Polonaise des Sciences et des Lettres. 1937;**35**:355-357

[15] Bender M, Dick G, Ge M, Deng Z, Wickert J, Kahle H-G, et al. Development of a GNSS water vapour

tomography system using algebraic reconstruction techniques. Advances in Space Research. 2011;**47**(10):1704-1720. DOI: 10.1016/j.asr.2010.05.034

[16] Stolle C. Three-dimensional imaging of ionospheric electron density fields using GPS observations at the ground and onboard the CHAMP satellite. [doctoral thesis]. Universität Leipzig: Institut für Meteorologie. 2014

[17] Jin S, Park JU. GPS ionospheric tomography: A comparison with the IRI-2001 model over South Korea. Earth, Planets and Space. 2007;**59**(4):287-292. DOI: 10.1186/BF03353106

[18] Gordon R, Bender R, Herman GT. Algebraic reconstruction technique (ART) for three-dimensional electron microscopy and X-ray photography. Journal of Theoretical Biology. 1970; **29**(3):471-481. DOI: 10.1016/0022-519 (370)90109-8

[19] Gilbert PFC. Iterative methods for three-dimensional reconstruction of an object from its projections. Journal of Theoretical Biology. 1972;**36**(1):105-117. DOI: 10.1016/0022-519(372)90180-4

[20] Moore EH. On the reciprocal of the general algebraic matrix. Bulletin of American Mathematical Society. 1920; **26**:394-395

[21] Penrose R. A generalized inverse for matrices. Proceedings of the Cambridge Philosophical Society. 1955;**51**(3): 406-413. DOI: 10.1017/ S0305004100030401

[22] Strang G, Borre K. Linear Algebra, Geodesy, and GPS. 1st ed. Wellesley, Massachusetts, USA: Wellesley-Cambridge Press; 1997. p. 624. ISBN: 978-0961408862

[23] Hansen PC. The L-curve and its use in the numerical treatment of inverse problems. In: Computational Inverse Problems in Electrocardiology. Vol. 4. Ashurst, UK: WIT Press; 2000. pp. 119-142

[24] Flores A. Atmospheric tomography using satellite radio signals [doctoral thesis]. Universitat Politecnica de Catalunya: Departament de Teoria del Senyal i Comunicacions. 1999

[25] Tikhonov AN. Solution of incorrectly formulated problems and the regularization method. Soviet Mathematics Doklady. 1963;**4**:1035-1038

[26] Elden L. Algorithms for the regularization of ill-conditioned least squares problems. BIT. 1977;**17**(2): 134-145. DOI: 10.1007/BF01932285

[27] WMO. Guide to Meteorological Instruments and Methods of Observation. 7th ed. Geneva, Switzerland: Secretariat of the World Meteorological Organization; 2008. ISBN: 978-9263100085

[28] Steiner AK, Löscher A, Kirchengast G. Error characteristics of refractivity profiles retrieved from CHAMP radio occultation data. In: Atmosphere and Climate. Berlin Heidelberg: Springer; 2006. pp. 27-36. DOI: 10.1007/3-540-34121-8

[29] Moeller G, Ao C, Mannucci T. Tomographic radio occultation methods applied to a dense cubesat formation in low Mars orbit. Radio Science. 2019; **56**(7):1-10. DOI: 10.1029/2020RS007199

[30] Fishbach FF. A satellite method for pressure and temperature below 24 km. Bulletin of the American Meteorological Society. 1965;**46**(9):528-532. DOI: 10.1175/1520-0477-46.9.528

[31] Phinney RA, Anderson DL. On the radio occultation method for study

planetary atmospheres. Journal of Geophysical Research. 1968;**73**(5): 1819-1827. DOI: 10.1029/ JA073i005p01819

[32] Rangaswamy S. Recovery of atmospheric parameters from the Apollo/Soyuz-ATS-F radio occultation data. Geophysical Research Letters. 1976; **3**(8):483-486. DOI: 10.1029/ GL003i008p00483

[33] Gorbunov ME. Three-dimensional satellite refractive tomography of the atmosphere: Numerical simulation. Radio Science. 1996;**31**(1):95-104. DOI: 10.1029/95RS01353

[34] Xu G. GPS Theory, Algorithms and Applications. 2nd ed. Berlin Heidelberg: Springer-Verlag; 2007. DOI: 10.1007/ 978-3-540-72715-6

[35] Svehla D. Geometrical Theory of Satellite Orbits and Gravity Field. 1st ed. Cham, Switzerland: Springer International Publishing; 2018. DOI: 10.1007/978-3-319-76873-1

[36] Kursinski ER, Hajj GA, Schofield JT, Linfield RP, Hardy KR. Observing earth's atmosphere with radio occultation measurements using the global positioning system. Journal of Geophysical Research. 1997;**102** (D19):23429-23465. DOI: 10.1029/ 97JD01569

[37] Aragon B, Houborg R, Tu K, Fisher JB, McCabe M. CubeSats enable high spatiotemporal retrievals of crop-water use for precision agriculture. Remote Sensing. 2018;**10**(12):1867. DOI: 10.3390/ rs10121867

[38] Douglas E, Cahoy KL, Morgan RE, Knapp M. CubeSats for astronomy and astrophysics. Bulletin of the AAS. 2019; **51**(7):1-6

[39] Curzi G, Modenini D, Tortora P. Large constellations of small satellites: A survey of near future challenges and missions. Aerospace. 2020;**7**(9):133. DOI: 10.3390/aerospace7090133

[40] de Carvalho RA, Estela J, Langer M. Nanosatellites, Nanosatellites: Space and Ground Technologies, Operations and Economics. Toronto, Canada: John Wiley & Sons; 2020. p. xxxv. ISBN: 978-1119042051

[41] Moeller G, Rothacher M, Sonnenberg F, Wolf A. A high-precision commercial off-the-shelf GNSS payload board for nanosatellite orbit determination and timing. Proceedings of the 44th COSPAR Scientific Assembly 16-24 July 2022, online

[42] Turk FJ, Padulles R, Ao CO, de la Torre JM, Wang KN, Franklin GW. Benefits of a closely-spaced satellite constellation of atmospheric polarimetric radio occultation measurements. Remote Sensing. 2019; **11**(20):1-19. DOI: 10.3390/rs11202399

Section 2

Some Computational Aspects

Chapter 4

Numerical Gradient Computation for Simultaneous Detection of Geometry and Spatial Random Fields in a Statistical Framework

Michael Conrad Koch, Kazunori Fujisawa and Akira Murakami

Abstract

The target of this chapter is the evaluation of gradients in inverse problems where spatial field parameters and geometry parameters are treated separately. Such an approach can be beneficial especially when the geometry needs to be detected accurately using L_2-norm-based regularization. Emphasis is laid upon the computation of the gradients directly from the governing equations. Working in a statistical framework, the Karhunen-Loève (K-L) expansion is used for discretization of the spatial random field and inversion is done using the gradient-based Hamiltonian Monte Carlo (HMC) algorithm. The HMC gradients involve sensitivities w.r.t the random spatial field and geometry parameters. Building on a method developed by the authors, a procedure is developed which considers the gradients of the associated integral eigenvalue problem (IEVP) as well as the interaction between the gradients w.r.t random spatial field parameters and the gradients w.r.t the geometry parameters. The same mesh and linear shape functions are used in the finite element method employed to solve the forward problem, the artificial elastic deformation problem and the IEVP. Analysis of the rate of convergence using seven different meshes of increasing density indicates a linear rate of convergence of the gradients of the log posterior.

Keywords: sensitivity analysis, geometry detection, random fields, Hamiltonian Monte Carlo, inverse problems

1. Introduction

Accurate computation of gradients, w.r.t parameters of interest, is a key aspect of deterministic algorithms like Gauss-Newton, Levenberg–Marquardt, Occam's inversion [1] as well as statistical algorithms like Hamiltonian Monte Carlo (HMC) [2]. Common nonintrusive methods like finite differences compute the gradient by taking differences between the response at the current model and at a perturbed model, such methods suffer from certain drawbacks. Two types of errors stand out in particular:

numerical error involved with truncation of Taylor's series and round-off error involved with finite precision arithmetic of computers [3]. For more robust analysis, this chapter focuses on methods where the gradient is computed directly from the analytical/numerical model by enlisting the sensitivity equations.

The type of solutions that can be obtained from inverse problems is guided by the regularization term. When the accurate detection of the geometry of embedded objects (or detection of discontinuities) is of interest, L_1-norm-based difference priors have proven to be useful [4, 5]. However, special techniques need to be used to accommodate the nondifferentiability of the L_1-norm [1]. Hence, to make use of a large library of adjoint-based inversion solvers (that use gradients w.r.t the parameters), L_2-norm-based priors which readily allow for differentiation are more popular in practice. However, L_2-norm regularization or Gaussian priors in a stochastic sense only admit smooth solutions. Hence, in order to still be able to capture discontinuities, this paper explicitly parameterizes the shape of the boundary (or the geometry of the domain). This approach thereby considers two sets of parameters, one related to the spatial field and the second with the geometry parameters. This approach is of course only applicable when the unknown geometry can be parameterized explicitly.

Gradients have to be computed w.r.t both spatial as well geometry parameters. While sensitivity analysis of spatial parameters is usually straightforward, computation of the gradients w.r.t geometry parameters [6, 7] needs to be done more carefully and consider aspects like mesh distortion. This chapter develops on the method presented in [8] and briefly details the simultaneous spatial field and geometry update within the HMC statistical framework. Similar to [8], the Karhunen-Loève (K-L) expansion is used for discretization of the random field, with the difference that the complete theoretical basis is considered. The complete integral eigenvalue problem or IEVP is solved and the associated gradients are computed. The interaction between the geometry and spatial parameters due to the domain of definition of the IEVP is also detailed. It should be noted that the procedure detailed above is applicable to both the direct differentiation and adjoint methods [9] of sensitivity analysis.

The entire numerical study is done with a focus on aspects related to gradients and not with the aim to solve the inverse problem. Nevertheless, a forward model is still required for the computation of gradients. The chapter begins with a description of the forward model, observation equation and the discretization of the spatial random field in Section 2. This is followed by a brief description in Section 3 of the HMC-based methods developed in [8, 10] for simultaneous spatial field and geometry detection. Section 3.3 introduces the new gradients obtained when the complete IEVP is considered. It is shown how the gradients of the eigenvectors involve the computation of a Moore-Penrose pseudoinverse. Finally, the gradient computation procedure is validated and a convergence study is done in Section 4.

2. Inversion preliminaries

2.1 Governing and observation equations

Consider a linear steady seepage flow problem defined on a domain $\mathbf{z} \in \Omega \subset \mathbb{R}^d$, $d \in \{2, 3\}$, where $\mathbf{k}(\mathbf{z})$ is a symmetric spatially varying hydraulic conductivity matrix, $h(\mathbf{z})$ is the hydraulic head, and $Q(\mathbf{z})$ is a source term as shown in Eq. (1) below

$$\nabla . -\mathbf{k}(\mathbf{z})\nabla h(\mathbf{z}) = Q(\mathbf{z}). \tag{1}$$

Standard Dirichlet: $h = \overline{h}$ on Γ_D, and Neumann: $f_n = \boldsymbol{f}.\boldsymbol{n} = -\mathbf{k}\nabla h.\boldsymbol{n}$ on Γ_N boundary conditions are applied. Following a spatial discretization of the weak form of the PDE using the finite element method, the governing equation can be written as:

$$\mathbf{K}(\boldsymbol{\theta})\mathbf{h} = \mathbf{q}, \tag{2}$$

where $\mathbf{K}$ is the global hydraulic conductivity matrix and $\mathbf{h}$ and $\mathbf{q}$ are the nodal hydraulic head and flux vectors respectively. The parameter vector $\boldsymbol{\theta} = ({}^1\boldsymbol{\theta}, {}^2\boldsymbol{\theta}) \in \mathbb{R}^K$ in Eq. (2), includes all the unknowns related to the inverse problem. In this study, the unknowns are divided into two sets [8]: ${}^1\boldsymbol{\theta} \in \mathbb{R}^{K_1}$, related to the spatial discretization of a random field and ${}^2\boldsymbol{\theta} \in \mathbb{R}^{K_2}$, and related to the definition of the geometry of the domain Ω. Also, consider an observation model relating the state vector $\mathbf{m} = (\mathbf{h}, \mathbf{q})$ to the discrete observations $\mathbf{y}$, through a map $\mathbf{H}$ that is independent of $\boldsymbol{\theta}$ i.e.

$$\mathbf{y} = \mathbf{Hm}(\boldsymbol{\theta}) + \mathbf{r}. \tag{3}$$

The error in Eq. (3) is modeled as Gaussian $\mathbf{r} \sim \mathbb{N}(\mathbf{0}, \mathbf{R})$ with a known covariance matrix $\mathbf{R}$. Parameter estimation is done in a probabilistic sense using Bayesian inference. Starting with a Gaussian prior distribution $p(\boldsymbol{\theta}) = \mathbb{N}(\boldsymbol{\theta}|\mathbf{0}, \boldsymbol{\Sigma}_{\boldsymbol{\theta}})$ and a likelihood distribution $p(\mathbf{y}|\boldsymbol{\theta}) = \mathbb{N}(\mathbf{y}|\mathbf{Hm}(\boldsymbol{\theta}), \mathbf{R})$, the posterior distribution is written as:

$$p(\boldsymbol{\theta}|\mathbf{y}) \propto p(\mathbf{y}|\boldsymbol{\theta})p(\boldsymbol{\theta}). \tag{4}$$

Except for linear Gaussian observation models, the posterior cannot be computed analytically and is usually evaluated using MCMC sampling algorithms.

2.2 Karhunen-Loève (K-L) expansion

The parameter vector ${}^1\boldsymbol{\theta}$ defined in Section 2.1 is associated with a continuous hydraulic conductivity spatial random field $k(\mathbf{z}, \omega)$, where $\mathbf{z}$ is defined on the domain Ω and ω belongs to the space of random events Θ. Let the expected value of the random field be denoted as $\overline{k} : \Omega \to \mathbb{R}$ and the autocovariance function $C : \Omega \times \Omega \to \mathbb{R}$ be defined as $C(\mathbf{z}, \mathbf{z}') = \sigma(\mathbf{z})\sigma(\mathbf{z}')\rho(\mathbf{z}, \mathbf{z}')$. Here $\sigma : \Omega \to \mathbb{R}$ is the standard deviation function and $\rho : \Omega \times \Omega \to [-1, 1]$ is the autocorrelation coefficient function. The study in this chapter is confined to Gaussian random fields that can be defined completely by their mean and autocovariance functions.

The Karhunen-Loève (K-L) expansion method is a series expansion method for the discretization of random fields which is based on the spectral decomposition of the autocovariance function. It can be shown that a random field can be written as an infinite sum [11]:

$$k(\mathbf{z}, \omega) = \overline{k}(\mathbf{z}) + \sum_{k=1}^{\infty} \sqrt{\lambda_k}\,{}^1\theta_k \phi_k(\mathbf{z}), \tag{5}$$

where ${}^1\theta_k : \Theta \to \mathbb{R}$ are standard uncorrelated random variables, λ_k are the eigenvalues (always non-negative) and $\phi_k(\mathbf{z})$ are the eigenfunctions of the linear operator related to the covariance kernel C. They can be obtained by solving the homogeneous Fredholm integral eigenvalue problem (IEVP) on the domain Ω:

$$\int_{\Omega} C(\mathbf{z}, \mathbf{z}')\phi_k(\mathbf{z}')\mathrm{d}\mathbf{z}' = \lambda_k \phi_k(\mathbf{z}). \tag{6}$$

The autocovariance function is symmetric, bounded, and positive semi-definite and has the spectral decomposition $C(\mathbf{z}, \mathbf{z}') = \sum_{k=1}^{\infty} \lambda_k \phi_k(\mathbf{z})\phi_k(\mathbf{z}')$. The eigenfunctions are orthogonal, and in a normalized form satisfy the condition $\int_{\Omega} \phi_k(\mathbf{z})\phi_l(\mathbf{z})\mathrm{d}\mathbf{z} = \delta_{kl}$, where δ_{kl} is the Kronecker delta. In the case of Gaussian random fields, the random variables ${}^1\theta_k$ are also independent and follow the standard normal distribution.

In practice, the eigenvalues decay exponentially fast for smooth functions and algebraically fast for non-smooth autocovariance kernels and the K-L expansion is usually truncated after K_1 terms. If the eigenvalues are arranged in descending order such that $\lambda_1 > \lambda_2 > \ldots > \lambda_{K_1}$, then accompanied by the associated eigenfunctions, the truncated K-L expansion approximation of the random field can be written as

$$\hat{k}(\mathbf{z}, \omega) = \overline{k}(\mathbf{z}) + \sum_{k=1}^{K_1} \sqrt{\lambda_k}\, {}^1\theta_k \phi_k(\mathbf{z}), \tag{7}$$

The truncated K-L expansion approximation is optimal in the sense that, for a fixed number of terms K_1, the mean square error over the domain is minimized [12]. A global error measure related to random field discretization is called the *mean error variance* $\overline{\varepsilon}_\sigma$ and is defined as [13]:

$$\overline{\varepsilon}_\sigma(\mathbf{z}) = \frac{1}{|\Omega|} \int_{\Omega} \frac{\mathrm{Var}\left[k(\mathbf{z}, \omega) - \hat{k}(\mathbf{z}, \omega)\right]}{\mathrm{Var}[k(\mathbf{z}, \omega)]} \mathrm{d}\mathbf{z}. \tag{8}$$

It can be shown that the variance of the truncated K-L expansion $\hat{k}(\mathbf{z}, \omega)$ is:

$$\mathrm{Var}\left[\hat{k}(\mathbf{z}, \omega)\right] = \sum_{k=1}^{K_1} \lambda_k \phi_k^2(\mathbf{z}). \tag{9}$$

Using the property $\mathbb{E}\left[{}^1\theta_k {}^1\theta_l\right] = \delta_{kl}$, the mean error variance can be calculated as [14]:

$$\overline{\varepsilon}_{\sigma,\mathrm{KL}} = 1 - \frac{1}{|\Omega|\sigma^2} \sum_{k=1}^{K_1} \lambda_k. \tag{10}$$

The derivation for Eq. (10) assumes the random field to be homogeneous, i.e., $\sigma(\mathbf{z}) = \sigma$. We only consider the case where the prior random field is homogeneous and Gaussian. This assumption is for numerical convenience and does not limit the posterior, which can be non-Gaussian and non-homogeneous [15].

2.3 Galerkin finite element method to solve the integral eigenvalue problem

The Galerkin Finite Element Method (FEM) is used to solve the IEVP on Ω. The eigenfunctions are approximated with the help of the shape functions $N_j : \Omega \to \mathbb{R}$ of the FE mesh, and is represented as:

$$\phi_k(\mathbf{z}) \approx \sum_{j=1}^{n} d_{kj} N_j(\mathbf{z}), \tag{11}$$

where the coefficients $d_{kj} \in \mathbb{R}$ are unknown and n is the number of nodes in the FE mesh. Substitution of Eq. (11) into Eq. (6), yields the residual:

$$r(\mathbf{z}) = \sum_{j=1}^{n} d_{kj} \left(\int_{\Omega} C(\mathbf{z}, \ \mathbf{z}') N_j(\mathbf{z}') \mathrm{d}\mathbf{z}' - \lambda_j N_j(\mathbf{z}) \right), \tag{12}$$

In the Galerkin method, the unknown coefficients are determined by making the residual $r(\mathbf{z})$ orthogonal to the space spanned by the shape functions i.e.

$$\int_{\Omega} r(\mathbf{z}) N_i(\mathbf{z}') \mathrm{d}\mathbf{z}' = 0 \forall j = 1, \dots, n. \tag{13}$$

This results in a generalized eigenvalue problem

$$\mathbf{B}\mathbf{d}_k = \lambda_k \mathbf{M}\mathbf{d}_k, \tag{14}$$

where

$$B_{ij} = \int_{\Omega} N_i(\mathbf{z}) \int_{\Omega} C(\mathbf{z}, \ \mathbf{z}') N_j(\mathbf{z}') \mathrm{d}\mathbf{z}' \mathrm{d}\mathbf{z} \text{ and}$$
$$M_{ij} = \int_{\Omega} N_i(\mathbf{z}) N_j(\mathbf{z}) \mathrm{d}\mathbf{z}. \tag{15}$$

Both $\mathbf{B}$ and $\mathbf{M}$ are $n \times n$ matrices that involve integrals over the domain Ω. Hence the actual geometry of the domain has to be considered for integration. The maximum number of available eigenpairs is n, but in practice, the K-L expansion can usually be truncated at K_1 terms such that $K_1 \ll n$. As such, for computational efficiency, it is sufficient to compute the first K_1 eigenpairs only, which can be done through the Lanczos algorithm.

3. Simultaneous geometry and spatial field detection

3.1 Hamiltonian Monte Carlo

Consider a parameter space $\boldsymbol{\theta} \in \mathbb{R}^K$ augmented with equidimensional momentum variables $\mathbf{p} \in \mathbb{R}^K$ and a joint probability distribution with density $p(\boldsymbol{\theta}, \mathbf{p})$ defined over this augmented space. If the underlying distribution over the momentum variables is chosen to be a Gaussian: $p(\mathbf{p}) \equiv \mathbb{N}(\mathbf{p}|\mathbf{0}, \ \mathcal{M})$, where $\mathcal{M}$ is user-specified and independent of $\boldsymbol{\theta}$, then a joint probability distribution can be defined as $p(\boldsymbol{\theta}, \mathbf{p}) = p(\mathbf{p})p(\boldsymbol{\theta}|\mathbf{y})$. The Hamiltonian $H : \mathbb{R}^K \times \mathbb{R}^K \to \mathbb{R}$ is then defined as:

$$H(\boldsymbol{\theta}, \mathbf{p}) = -\log p(\mathbf{p}) - \log p(\boldsymbol{\theta}|\mathbf{y}) \tag{16}$$

The second term on the right of Eq. (17) is generally called $\varphi : \mathbb{R}^K \to \mathbb{R}$ and is given as $\varphi(\boldsymbol{\theta}) = -\log p(\boldsymbol{\theta}|\mathbf{y})$.

The introduction of the momentum variables allows for the generation of trajectories through conservative Hamiltonian dynamics [16], which are given as:

$$\begin{pmatrix} \frac{d\boldsymbol{\theta}}{dt} \\ \frac{d\mathbf{p}}{dt} \end{pmatrix} = \begin{bmatrix} 0 & 1 \\ -1 & 0 \end{bmatrix} \begin{pmatrix} \frac{\partial H}{\partial \boldsymbol{\theta}} \\ \frac{\partial H}{\partial \mathbf{p}} \end{pmatrix} = \begin{pmatrix} -\frac{\partial \varphi(\boldsymbol{\theta})}{\partial \boldsymbol{\theta}} \\ \mathcal{M}^{-1}\mathbf{p} \end{pmatrix}. \tag{17}$$

These dynamics are exactly reversible (provided the gradient $\frac{\partial \varphi(\boldsymbol{\theta})}{\partial \boldsymbol{\theta}}$ is one-to-one) and preserve volume as Eq. (18) is just a rotation transformation in $\boldsymbol{\theta} - \mathbf{p}$ space. Except for simple problems Eq. (18) cannot be solved analytically and is usually solved using the leapfrog method, which is a second-order accurate numerical integrator given as:

$$\mathbf{p}\left(t + \frac{\epsilon}{2}\right) = \mathbf{p}(t) - \frac{\epsilon}{2} \frac{\partial \varphi(\boldsymbol{\theta}(t))}{\partial \boldsymbol{\theta}}, \tag{18}$$

$$\boldsymbol{\theta}(t + \epsilon) = \boldsymbol{\theta}(t) + \epsilon \mathcal{M}^{-1}\mathbf{p}\left(t + \frac{\epsilon}{2}\right) \text{ and} \tag{19}$$

$$\mathbf{p}(t + \epsilon) = \mathbf{p}\left(t + \frac{\epsilon}{2}\right) - \frac{\epsilon}{2} \frac{\partial \varphi(\boldsymbol{\theta}(t + \varepsilon))}{\partial \boldsymbol{\theta}}. \tag{20}$$

Starting from a point $(\boldsymbol{\theta}^j, \mathbf{p}^j)$, these equations are applied repeatedly for L steps, each with a step-size ϵ, to determine a transition to a new point $(\boldsymbol{\theta}^{j+1}, \mathbf{p}^{j+1})$, which lies on the same Hamiltonian level-set as $(\boldsymbol{\theta}^j, \mathbf{p}^j)$. The deterministic part of Hamiltonian Monte Carlo (HMC) [2] is defined by Eqs. (19)–(21). The stochastic part of HMC comes from resampling $\mathbf{p} \sim \mathbb{N}(\mathbf{0}, \mathcal{M})$. The statistical efficiency of Hamiltonian Monte Carlo stems from the fact that the gradient-guided transitions can propose new points that are "far-away" from the starting point, thereby enabling efficient sampling of the posterior. This is in contrast to the random nature of transitions, which suffer from the curse of dimensionality [17], in conventional MCMC algorithms. As shown in Eqs. (19)–(21), critical to the success of HMC, is the computation of the gradient $\frac{\partial \varphi(\boldsymbol{\theta})}{\partial \boldsymbol{\theta}}$. Special attention must be paid to maintaining the reversibility of the transitions to satisfy the detailed balance condition [18] for MCMC algorithms. This is detailed along with the gradient computation procedure in the following sections.

3.2 Parameter update using the mesh moving method

The leapfrog equations determine an update in $\boldsymbol{\theta} - \mathbf{p}$ space. In particular, the update from $\boldsymbol{\theta}(t) \to \boldsymbol{\theta}(t + \epsilon)$ in Eq. (20), defines not only a new realization of the random field, but also a new domain, i.e., $\Omega(^2\boldsymbol{\theta})$. Without loss of generality, consider a domain in 2D discretized such that $\mathbf{Z}(^2\boldsymbol{\theta}) \in \mathbb{R}^2$ is the nodal coordinate vector of all the nodes of the finite element mesh. Let $\mathbf{Z}_v(^2\boldsymbol{\theta}) \in \mathbb{R}^2$ represent a subset of this vector that includes only the node coordinates at the piping zone boundary. Koch et al. [10] show that the computation of the gradient $\frac{\partial \varphi(\boldsymbol{\theta})}{\partial \boldsymbol{\theta}}$, by analytical methods [6], ultimately involves the computation of the gradient of the nodal coordinate vector $\frac{\partial \mathbf{Z}(\boldsymbol{\theta})}{\partial \boldsymbol{\theta}}$.

Computation of the nodal coordinate vector gradient requires the definition of a differentiable map which is additionally reversible and one-to-one, to satisfy the detailed balance condition of MCMC. One such map proposed in [10], is to update from an arbitrarily fixed reference domain $\Omega^{ref}(^2\boldsymbol{\theta}^{ref})$, defined by an arbitrary parameter $^2\boldsymbol{\theta}^{ref}$. Let $\mathbf{Z}^{ref}(^2\boldsymbol{\theta}^{ref}) \in \mathbb{R}^{n\times 2}$ be the nodal coordinate vector on the discretized domain Ω^{ref} and $\mathbf{Z}_v^{ref}(^2\boldsymbol{\theta}^{ref}) \in \mathbb{R}^{n_v\times 2}$ represent a subset of this vector that includes only the coordinates of the n_v nodes at the piping zone boundary. The nodal coordinates $\mathbf{Z}_v(^2\boldsymbol{\theta})$ and $\mathbf{Z}_v^{ref}(^2\boldsymbol{\theta}^{ref})$ can be determined explicitly if $^2\boldsymbol{\theta}$ and $^2\boldsymbol{\theta}^{ref}$ are known respectively. Following the update, $^2\boldsymbol{\theta}(t) \rightarrow {}^2\boldsymbol{\theta}(t+\varepsilon)$, $\mathbf{Z}_v(^2\boldsymbol{\theta}(t+\varepsilon))$ is available, and an artificial elastic deformation problem can be set up from the arbitrary known reference domain $\Omega^{ref}(^2\boldsymbol{\theta}^{ref})$ to the current domain $\Omega(^2\boldsymbol{\theta}(t+\varepsilon))$. The prescribed displacements $\mathbf{u}_v^{ref} \in \mathbb{R}^{n_v\times 2}$ for the elastic deformation problem are given as:

$$\mathbf{u}_v^{ref} = \mathbf{Z}_v(^2\theta(t+\varepsilon)) - \mathbf{Z}_v^{ref}(^2\theta^{ref}). \tag{21}$$

The entire mesh is moved and the new nodal coordinates can be determined as:

$$\mathbf{Z}(^2\boldsymbol{\theta}(t+\varepsilon)) = \mathbf{Z}^{ref}(^2\boldsymbol{\theta}^{ref}) + \mathbf{u}^{ref}, \tag{22}$$

where $\mathbf{u}^{ref} \in \mathbb{R}^{n\times 2}$ represents the displacement of all the nodes from the reference domain to the current domain.

The displacements in the elastic deformation step can cause distortions in the mesh, especially in regions where large deformation is expected, i.e., near the piping zone boundary. To maintain a good mesh quality for computation purposes, a mesh moving method [19] is used. The simple idea is to scale the elastic modulus E_e^{ref} of each element (in the reference domain) with the determinant of the Jacobian $|\mathbf{J}_e^{ref}|$ in the elastic deformation step:

$$\tilde{E}_e^{ref} = E_e^{ref}\left(1/|\mathbf{J}_e^{ref}|\right)^{\chi^{ref}}, \tag{23}$$

where χ^{ref} is an arbitrary positive scaling parameter. The Poisson's ratio of the reference domain ν^{ref} is also chosen arbitrarily. The performance of the algorithm has been shown [20] to be invariant to the choice of these reference parameters. The net effect of such scaling is that small elements become rigid and larger elements become more flexible. Hence, if the reference domain mesh is constructed carefully such that small elements are placed in regions where large distortion is expected, i.e., near the piping zone, and larger elements are placed in regions of less expected distortion, the method is expected to yield a good mesh quality in the elastic deformation stage. The map in Eq. (23) is differentiable and helps determine $\frac{\partial \mathbf{Z}(\boldsymbol{\theta})}{\partial \boldsymbol{\theta}}$.

3.3 Gradient computation

Analytical methods for the computation of the gradient are intrusive and involve gradients of the steady seepage flow forward solver. From the definition of $\varphi(\boldsymbol{\theta}) = -\log p(\boldsymbol{\theta}|\mathbf{y})$, it is apparent that the computation of $\frac{\partial \varphi(\boldsymbol{\theta})}{\partial \boldsymbol{\theta}}$ requires the computation of $\frac{\partial \mathbf{m}}{\partial \boldsymbol{\theta}} = \left(\frac{\partial \mathbf{h}}{\partial \boldsymbol{\theta}}, \frac{\partial \mathbf{q}}{\partial \boldsymbol{\theta}}\right)$. These terms can be computed by taking the derivative of Eq. (2) and is given as:

$$\frac{\partial \mathbf{K}}{\partial \boldsymbol{\theta}}\mathbf{h} + \mathbf{K}\frac{\partial \mathbf{h}}{\partial \boldsymbol{\theta}} = \frac{\partial \mathbf{q}}{\partial \boldsymbol{\theta}}. \tag{24}$$

Given standard boundary conditions, this equation can be solved to obtain $\frac{\partial \mathbf{h}}{\partial \boldsymbol{\theta}}$ and $\frac{\partial \mathbf{q}}{\partial \boldsymbol{\theta}}$, provided $\frac{\partial \mathbf{K}}{\partial \boldsymbol{\theta}}$ is known.

Following the standard procedure in FEM, the hydraulic conductivity matrix at the element level Ω_e in the current domain is given as:

$$\mathbf{K}_e = \int_{\Omega_e} \mathbf{G}^{\mathrm{T}}\hat{k}(\boldsymbol{\theta}, \mathbf{z})\mathbf{G}|\mathbf{J}_e|\mathrm{d}\boldsymbol{\xi}, \tag{25}$$

where $\hat{k}$ represents the hydraulic conductivity spatial field obtained from the truncated K-L expansion, $\mathbf{G}$ contains the derivatives (w.r.t $\mathbf{z}$) of the shape functions N_j described earlier in Section 2.3, $|\mathbf{J}_e|$ is the determinant of the Jacobian matrix associated with the isoparametric transformation $(\xi_1, \xi_2) \rightarrow (z_1, z_2)$ and Ω_e is the region occupied by the parent element related to the isoparametric transformation. The gradient of $\mathbf{K}_e$ with respect to the spatial parameters ${}^1\boldsymbol{\theta}$ can be computed as:

$$\frac{\partial \mathbf{K}_e}{\partial^1 \boldsymbol{\theta}} = \int_{\Omega_e} \mathbf{G}^{\mathrm{T}}\frac{\partial \hat{k}}{\partial^1 \boldsymbol{\theta}}\mathbf{G}|\mathbf{J}_e|\mathrm{d}\boldsymbol{\xi}, \tag{26}$$

where the gradient of the hydraulic conductivity field w.r.t ${}^1\boldsymbol{\theta}$ is easily obtained by differentiating Eq. (7) and is given as:

$$\frac{\partial \hat{k}}{\partial^1 \theta_j} = \sqrt{\lambda_j}\phi_j, \tag{27}$$

The gradient of $\mathbf{K}_e$ w.r.t the geometry parameters ${}^2\boldsymbol{\theta}$ can be written as [6]:

$$\frac{\partial \mathbf{K}_e}{\partial^2 \boldsymbol{\theta}} = \int_{\Omega_e}\left(\frac{\partial \mathbf{G}^{\mathrm{T}}}{\partial^2 \boldsymbol{\theta}}\hat{k}\mathbf{G}|\mathbf{J}_\mathrm{e}| + \mathbf{G}^{\mathrm{T}}\hat{k}\frac{\partial \mathbf{G}}{\partial^2 \boldsymbol{\theta}}|\mathbf{J}_\mathrm{e}| + \mathbf{G}^{\mathrm{T}}\hat{k}\mathbf{G}\frac{\partial|\mathbf{J}_\mathrm{e}|}{\partial^2 \boldsymbol{\theta}} + \mathbf{G}^{\mathrm{T}}\frac{\partial \hat{k}}{\partial^2 \boldsymbol{\theta}}\mathbf{G}|\mathbf{J}_\mathrm{e}|\right)\mathrm{d}\boldsymbol{\xi}. \tag{28}$$

Formulas for the calculation of the gradients $\frac{\partial \mathbf{G}}{\partial^2 \boldsymbol{\theta}}$ and $\frac{\partial |\mathbf{J}_\mathrm{e}|}{\partial^2 \boldsymbol{\theta}}$ can readily be found in literature [6]. It is clear from Eq. (6) that the computation of the eigenvalues and eigenfunctions depends on the definition of the domain $\Omega({}^2\boldsymbol{\theta})$. Hence, the gradient $\frac{\partial \hat{k}(\boldsymbol{\theta}, \mathbf{z})}{\partial^2 \boldsymbol{\theta}}$ will involve the gradients of the eigenvalues and eigenvectors and is written as:

$$\frac{\partial \hat{k}}{\partial^2 \boldsymbol{\theta}} = \sum_{j=1}^{K_1} \left(\frac{1}{2\sqrt{\lambda_j}} \frac{\partial \lambda_j}{\partial^2 \boldsymbol{\theta}} \phi_j + \sqrt{\lambda_q} \frac{\partial \phi_j}{\partial^2 \boldsymbol{\theta}} \right) {}^1\theta_j. \tag{29}$$

The gradient of the eigenvalues and eigenvectors of the generalized eigenvalue problem in Eq. (14) w.r.t. $^2\boldsymbol{\theta}$ are written as [21]:

$$\frac{\partial \lambda_j}{\partial^2 \boldsymbol{\theta}} = \mathbf{d}_j^{\mathrm{T}} \left(\frac{\partial \mathbf{B}}{\partial^2 \boldsymbol{\theta}} - \lambda_j \frac{\partial \mathbf{M}}{\partial^2 \boldsymbol{\theta}} \right) \mathbf{d}_j, \tag{30}$$

$$\frac{\partial \mathbf{d}_j}{\partial^2 \boldsymbol{\theta}} = \left(\lambda_j \mathbf{M} - \mathbf{B} \right)^{\dagger} \left(\frac{\partial \mathbf{B}}{\partial^2 \boldsymbol{\theta}} - \lambda_j \frac{\partial \mathbf{M}}{\partial^2 \boldsymbol{\theta}} \right) \mathbf{d}_j - \frac{1}{2} \left(\mathbf{d}_j^{\mathrm{T}} \frac{\partial \mathbf{M}}{\partial^2 \boldsymbol{\theta}} \mathbf{d}_j \right) \mathbf{d}_j. \tag{31}$$

As $\mathbf{d}_j$ lies in the null space of $\lambda_j \mathbf{M} - \mathbf{B}$, the inverse $\left(\lambda_j \mathbf{M} - \mathbf{B} \right)^{-1}$ cannot be computed and a generalized inverse called the Moore-Penrose pseudoinverse $(\bullet)^{\dagger}$ is employed. In this study, the Moore-Penrose pseudoinverse is calculated by first carrying out an SVD of the matrix $\lambda_j \mathbf{M} - \mathbf{B}$ and then eliminating the smallest singular values below a tolerance level. This is then followed by taking a standard inverse of the SVD.

Considering the same mesh that is used for the solution of the steady seepage flow problem to be used for the discretization of the random field, the gradients of the matricces $\mathbf{B}$ and $\mathbf{M}$ at an elemental level in isoparametric space are given as:

$$\begin{aligned} \frac{\partial B_{ij}}{\partial^2 \boldsymbol{\theta}} &= \int_{\Omega_e} N_i \int_{\Omega_e} C(\mathbf{z}, \mathbf{z}') N_j \left(\frac{\partial |\mathbf{J}_e(\mathbf{z}')|}{\partial^2 \boldsymbol{\theta}}, |\mathbf{J}_e(\mathbf{z}')| + |\mathbf{J}_e(\mathbf{z})| \frac{\partial |\mathbf{J}_e(\mathbf{z}')|}{\partial^2 \boldsymbol{\theta}} \right) \mathrm{d}\xi' \mathrm{d}\xi \\ &+ \int_{\Omega_e} N_i \int_{\Omega_e} \left(\frac{\partial C(\mathbf{z}, \mathbf{z}')}{\partial^2 \boldsymbol{\theta}} \right) N_j |\mathbf{J}_e(\mathbf{z})| |\mathbf{J}_e(\mathbf{z}')| \mathrm{d}\xi' \mathrm{d}\xi, \end{aligned} \tag{32}$$

$$\frac{\partial M_{ij}}{\partial^2 \boldsymbol{\theta}} = \int_{\Omega_e} N_i N_j \frac{\partial |\mathbf{J}_e(\mathbf{z}))|}{\partial^2 \boldsymbol{\theta}} \mathrm{d}\xi. \tag{33}$$

In this study, the squared exponential autocorrelation coefficient function has been used, i.e.,

$$\rho(\mathbf{z}, \mathbf{z}') = \exp\left(\frac{-|\mathbf{z} - \mathbf{z}'|^2}{l_c^2} \right), \tag{34}$$

where l_c is the correlation length of the random field. Assuming $\sigma(\mathbf{z}) = \sigma(\mathbf{z}') = \sigma$, the gradient of the autocovariance function $C(\mathbf{z}, \mathbf{z}')$ w.r.t $^2\boldsymbol{\theta}$ is given as:

$$\frac{\partial C(\mathbf{z}, \mathbf{z}')}{\partial^2 \boldsymbol{\theta}} = -\frac{2}{l_c^2} \left(\left(\frac{\partial \mathbf{z}}{\partial^2 \boldsymbol{\theta}} - \frac{\partial \mathbf{z}}{\partial^2 \boldsymbol{\theta}} \right)^{\mathrm{T}} (\mathbf{z} - \mathbf{z}') \right) C(\mathbf{z}, \mathbf{z}'). \tag{35}$$

This completes the definition of all the terms required to compute the HMC gradient

$$\frac{\partial \varphi(\boldsymbol{\theta})}{\partial \boldsymbol{\theta}}.$$

4. Numerical implementation and results

4.1 Observation data

A seepage zone containing a piping region of length l and width w, shown in **Figure 1**, is chosen to generate synthetic observations for inversion. A steady seepage flow problem is solved on the domain defined by $l = 0.15$ m and $w = 0.05$ m. The left and right boundaries marked in blue are Dirichlet boundaries and the top and bottom boundaries are no-flow Neumann boundaries. Linear shape functions are used in the FE solution of the forward problem. Ten sets of observations (hydraulic head and total outward normal flux) are made by increasing the hydraulic head at the left boundary as shown by point A in **Figure 2**. The hydraulic head at the right boundary is fixed at 0. The corresponding hydraulic head recorded at observation points B, C, D, and E are shown in **Figure 2**. The total outward normal flux from the right boundary is summed and also shown on the right in **Figure 2**. Observation data is generated assuming a constant hydraulic conductivity field, i.e., $k^{\text{true}}(\mathbf{z}) = 0.001$ m/s. The standard deviation of the observation noise for the hydraulic head data and outward normal flux data is taken to be 1 and 5% respectively.

4.2 Inversion setup

Considering a log-normal hydraulic conductivity field $k(\mathbf{z})$, inversion is carried out on the Gaussian field $\tilde{k}(\mathbf{z}) = \log\left[k(\mathbf{z}) - \check{k}\right]$, where $\check{k}$ is a lower bound taken as 10^{-5} m/s. The log-normal construction enables a convenient choice of the standard deviation of the random field which is chosen as $\sigma(\mathbf{z}) = 1$. The correlation length of the

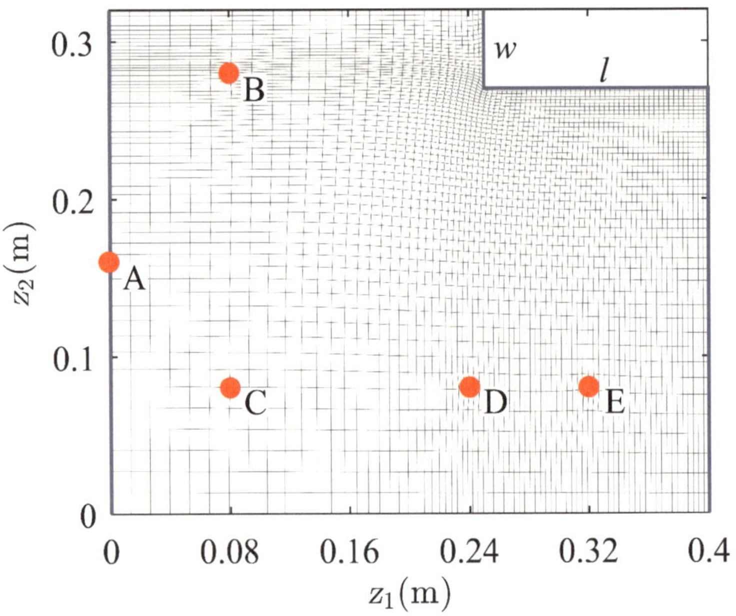

Figure 1.
Discretized seepage domain containing piping zone, used to obtain observation data, i.e., hydraulic head h data at points B, C, D and E, and total outward normal flux data q from the boundary marked in blue on the right. The hydraulic head is constant on the left boundary. The number of nodes in the discretization is 3943.

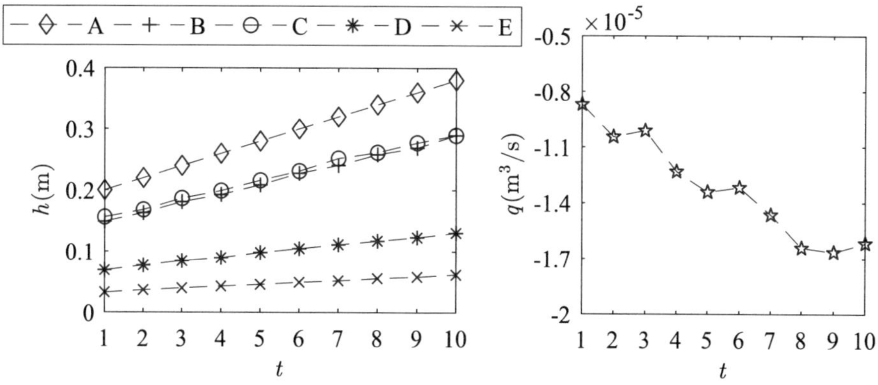

Figure 2.
Observation data generated from numerical seepage flow experiment. Data at each time t is obtained by solving a new seepage flow experiment with an input hydraulic head at the left boundary represented by A.

random field is assumed to be known and is taken to be equal to the length of the largest dimension of the domain i.e. $l_c = 0.4$m.

A preliminary assessment of the eigenvalues obtained by solving the generalized eigenvalue problem in Eq. (14) reveals that the K-L expansion can be truncated at $K_1 = 12$ terms. The eigenvalues are found to decay by approximately 3×10^4 times. This enables the determination of the number of terms in the spatial parameter vector ${}^1\boldsymbol{\theta} = (\theta_1, \ldots, \theta_{12})$. As mentioned in Section 3.2, once the geometry parameterization ${}^2\boldsymbol{\theta} = \left({}^2\theta_1, {}^2\theta_2\right) = (\theta_{13}, \theta_{14})$ is known (see **Figure 3(a)**), an explicit function $\mathbf{Z}_v\left({}^2\boldsymbol{\theta}\right)$ can be constructed as:

$$\mathbf{z}^1_{v_j} = \begin{pmatrix} L_1 - {}^2\theta_1 + (j-1)\dfrac{{}^2\theta_1}{n_1 - 1} \\ L_2 - {}^2\theta_2 \end{pmatrix}, \tag{36}$$

$$\mathbf{z}^2_{v_j} = \begin{pmatrix} L_1 - {}^2\theta_1 \\ L_2 - {}^2\theta_2 + (j-1)\dfrac{{}^2\theta_2}{n_2 - 1} \end{pmatrix}. \tag{37}$$

Eqs. (37) and (38) can be readily differentiated to obtain $\frac{\partial \mathbf{Z}_v}{\partial \boldsymbol{\theta}}$, which can be used to compute other gradients $\frac{\partial \mathbf{Z}}{\partial \boldsymbol{\theta}}, \frac{\partial \varphi}{\partial \boldsymbol{\theta}}$ etc. This completes the definition of the parameter vector.

As all updates are designed to take place from an arbitrary reference domain as mentioned in Eq. (23), the study of the convergence properties requires a focus on the reference mesh. The reference domain is arbitrarily chosen as the true domain used to generate the observations. Seven different reference configuration meshes with 128, 306, 497, 1038, 1476, 1860, and 3018 nodes are considered for the study of the convergence properties. Three of these meshes are shown in **Figure 3**. The artificial elastic properties selected for the mesh moving method are $E^{ref} = 25$ MPa, $\nu^{ref} = 0.25$ and $\chi^{ref} = 1$. To reduce mesh distortion during the mesh moving stage, all the reference meshes have a unique construction, i.e., smaller elements (which behave rigidly) are placed close to the piping zone boundary and larger elements (which are more flexible) are placed further away.

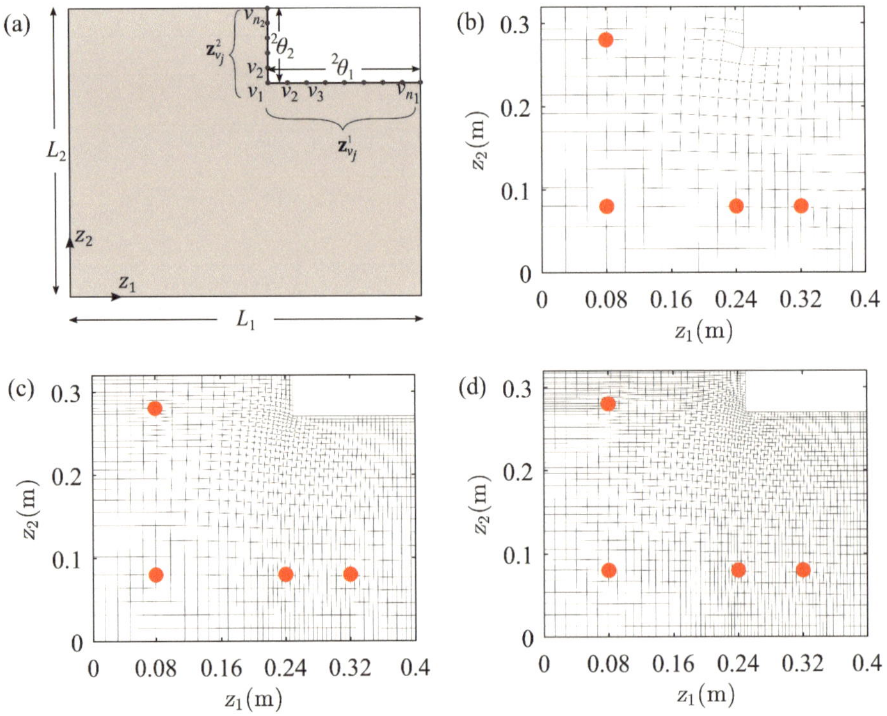

Figure 3.
(a) Parameterizations of the piping zone boundary. Discretization of the reference domain defined by $\left({}^2\theta_1^{ref}, {}^2\theta_2^{ref}\right) = (0.15\text{ m}, 0.05\text{ m})$ *with (b) 307 (c) 1476 and (d) 3018 nodes. Red dots indicate the position of observation points.*

4.3 Inversion and convergence analysis

There are no analytical solutions to verify the correctness of the gradient computation procedure mentioned in Section 3. Hence, for verification purposes, a short inversion analysis is done using HMC, where $i = 150$ samples are drawn from the posterior. The number of leapfrog steps is variable and drawn from a Gaussian $L \sim \mathbb{N}(5, 2)$. The prior for the parameters are chosen as ${}^1\boldsymbol{\theta} \sim \mathrm{N}(\mathbf{0}, \mathbf{I}_{12})$ and ${}^2\boldsymbol{\theta} \sim \mathrm{N}(\mathbf{0}, 0.1\mathbf{I}_2)$. The results presented in **Figure 4** correspond to inversion done considering the reference mesh shown in **Figure 3(b)**. The potential energy φ decreases rapidly in the first 20 steps and thereafter HMC begins the exploration of the region of the high posterior probability. Prior experience leads the authors to consider the performance of HMC to be appropriate and as such, is an indirect indicator that the gradient computation procedure is correct.

In order to study the convergence properties, suitable error measures are required. The rate of convergence of the truncated KL expansion of the prior random field, for a fixed number of parameters $K_1 = 12$, is determined through the relative mean error variance [14]:

$$\varepsilon_{\mathrm{Var,rel}} = \frac{\left|\overline{\varepsilon}_{\sigma,\mathrm{KL}} - \overline{\varepsilon}_{\sigma,\mathrm{analytic}}\right|}{\overline{\varepsilon}_{\sigma,\mathrm{analytic}}}. \tag{38}$$

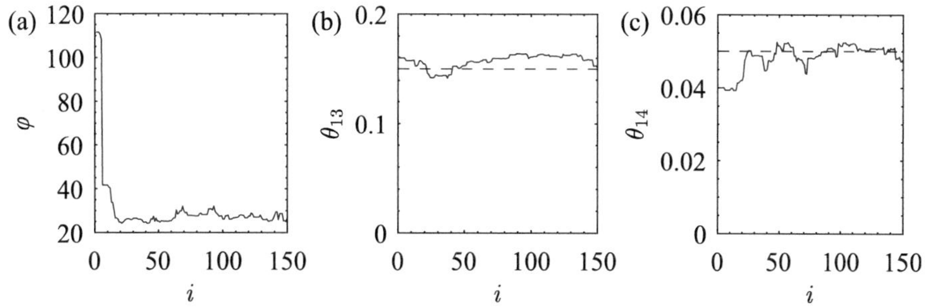

Figure 4.
Results from HMC showing (a) Decrease in potential energy as HMC progresses towards the high posterior probability region and samples of (b) θ_{13} and (c) θ_{14}. The dashed lines represent the true values of the parameters used to generate the observations.

The analytical mean error variance $\bar{\varepsilon}_{\sigma,\text{analytic}}$ cannot be computed for squared exponential type autocorrelation coefficient functions (Eq. (35)) and is calculated numerically using the fine mesh containing 3943 nodes, as shown in **Figure 1**. The relative error is computed for the seven different meshes mentioned in Section 4.2 and plotted in a log–log plot in **Figure 5**. A closer look at Eq. (10) indicates that the mean error variance is dependent only on the cumulative sum of the K_1 eigenvalues. As the eigenvalues are arranged in descending order, the relative error of the largest eigenvalue (λ_1) is also shown in **Figure 5**. The corresponding relative error measure can be written in a generalized manner as:

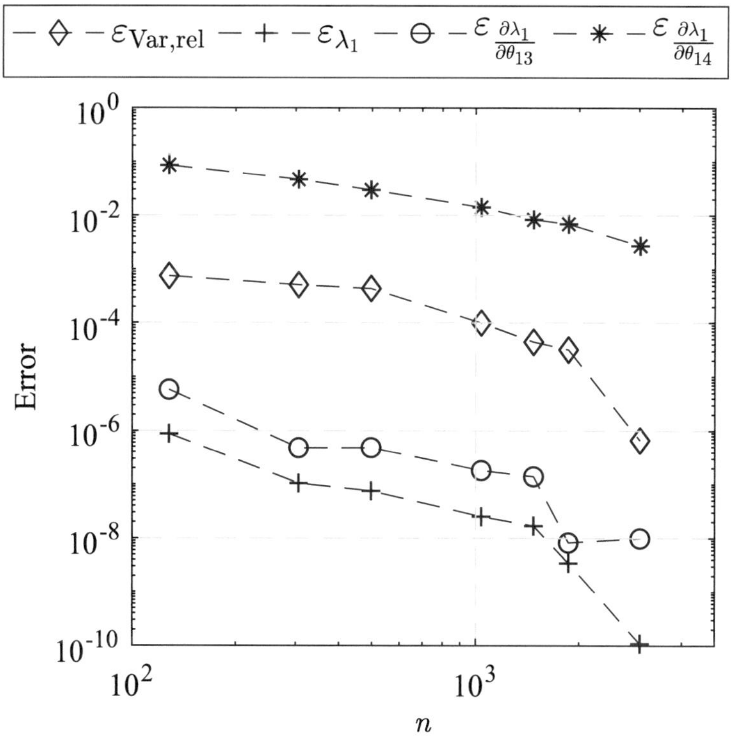

Figure 5.
Convergence of relative errors for different variables for a fixed number of terms $K_1 = 12$ in the K-L expansion.

$$\varepsilon_{\lambda_1} = \frac{|\lambda_1 - \lambda_{1\text{analytic}}|}{\lambda_{1\text{analytic}}}. \tag{39}$$

Other error measures can be computed in a similar manner just by replacing λ_1 with the relevant variable. Finally, the relative error of the gradient of the largest eigenvalue w.r.t the geometry parameters $\frac{\partial \lambda_1}{\partial^2 \boldsymbol{\theta}}$ is also shown in **Figure 5**. To compute the error related to the gradient, one mesh moving step is carried out, where the mesh is moved from the reference domain $\left({}^2\theta_1^{ref}, {}^2\theta_2^{ref}\right) = (0.15 \text{ m}, 0.05 \text{ m})$ to an arbitrary domain defined by $\left({}^2\theta_1, {}^2\theta_2\right) = (0.16 \text{ m}, 0.04 \text{ m})$. The same linear shape functions are used for discretization of the random field, the mesh moving method and the solution of the forward problem. The same realization of the parameter vector ${}^1\boldsymbol{\theta}$ is used to generate the random field in all the cases. A least squares fit of the relative error plots with a first-order polynomial reveals a linear convergence rate between −2.5 and − 5. Linear rates of convergence for $\varepsilon_{\text{Var,rel}}$, using linear FEM, are also observed by Betz et al. [14].

Finally, the rate of convergence of the gradient of the potential energy function w.r.t the spatial parameter related to the largest eigenvalue θ_1, and the two geometry parameters θ_{13} and θ_{14} is plotted in **Figure 6**. The convergence rates of the different gradients are almost parallel to each other. Once again, the slope of each plot is

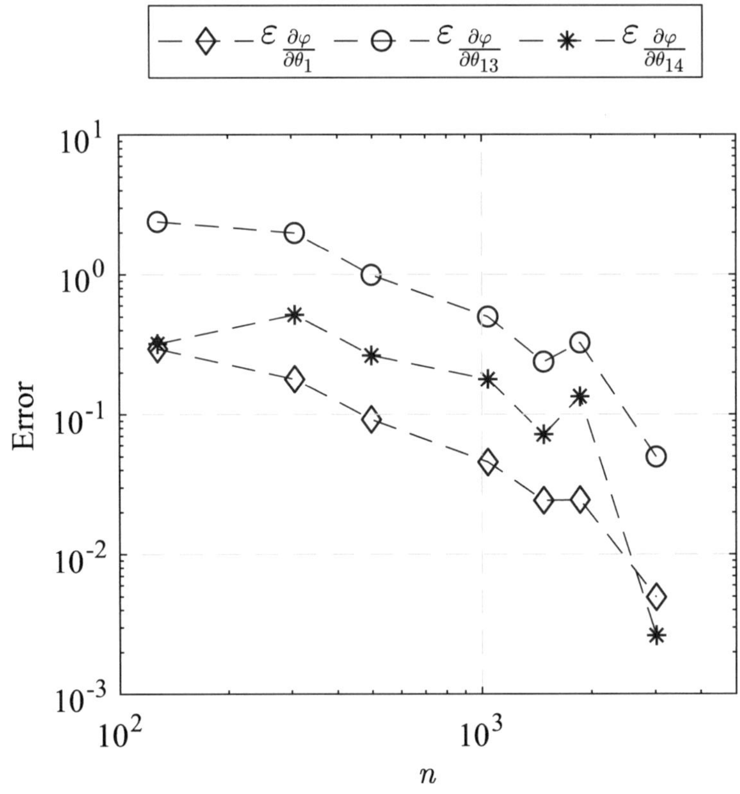

Figure 6.
Convergence of relative errors of potential energy gradients for a fixed number of terms $K_1 = 12$ in the K-L expansion.

determined by a least squares fit with a first-order polynomial. Each plot shows a linear convergence rate with a slope of approximately −2.7.

5. Conclusions

This paper details the numerical procedure for the computation of gradients in probabilistic inverse problems involving the simultaneous estimation of spatial fields and geometry. The method is analytical (in the sense of [6]), intrusive and involves the computation of gradients of the forward problem. Emphasis is laid upon the calculation of gradient of eigenvalues and eigenvectors involved with the truncated K-L expansion method for discretization of the random field. The eigenvalues and eigenvectors are obtained by solving a generalized eigenvalue problem on a defined domain. This implies that as the geometry parameters are updated, the domain is updated and the generalized eigenvalues and eigenvectors change. Computation of the gradient of the eigenvectors w.r.t the geometry parameters involve the computation of a generalized inverse.

The gradients are validated through an inverse analysis using HMC. The potential energy decreases rapidly as the Markov chain related to the geometry parameters approaches the region of high posterior probability, indicating the correctness of the computed gradients. Overall the rate of convergence of various quantities is observed to be linear on a log–log plot. This means computation costs can grow exponentially to achieve better results. While the same mesh that is used to solve the forward problem can also be used for the Galerkin FE method-based discretization of the IEVP, the repeated need to solve the forward problem, the mesh moving method and the generalized eigenvalue problem at every step can be computationally prohibitive. Elimination of any one, two or even all three of these numerical problems can significantly improve the computational feasibility of simultaneous spatial field and geometry detection using gradient-based stochastic samplers like HMC.

Acknowledgements

This work was supported by JSPS KAKENHI Grant Number JP21H02304.

Conflict of interest

The authors declare no conflict of interest.

Author details

Michael Conrad Koch*, Kazunori Fujisawa and Akira Murakami
Graduate School of Agriculture, Kyoto University, Kyoto, Japan

*Address all correspondence to: koch.michaelconrad.5w@kyoto-u.ac.jp

References

[1] Aster RC, Borchers B, Thurber CH. Parameter Estimation and Inverse Problems. Oxford: Academic Press; 2013. DOI: 10.1016/C2009-0-61134-X

[2] Neal R. MCMC Using Hamiltonian Dynamics. Handb. Markov Chain Monte Carlo. New York: CRC Press; 2011. DOI: 10.1201/b10905-6

[3] Voorhees A, Millwater H, Bagley R. Complex variable methods for shape sensitivity of finite element models. Finite Elements in Analysis and Design. 2011;**47**:1146-1156. DOI: 10.1016/J.FINEL.2011.05.003

[4] Lee J, Kitanidis PK. Bayesian inversion with total variation prior for discrete geologic structure identification. Water Resources Research. 2013;**49**: 7658-7669. DOI: 10.1002/2012WR013431

[5] Rudin LI, Osher S, Fatemi E. Nonlinear total variation based noise removal algorithms. Physica D: Nonlinear Phenomena. 1992;**60**: 259-268. DOI: 10.1016/0167-2789(92)90242-F

[6] Christensen PW, Klarbring A. Two-Dimensional Shape Optimization. An Introd. to Struct. Optim. Netherlands: Springer; 2008. DOI: 10.1007/978-1-4020-8666-3_7

[7] Haslinger J, Mäkinen RA. Introduction to Shape Optimization-Theory, Approximation, and Computation. Philadelphia: SIAM; 2003

[8] Koch MC, Osugi M, Fujisawa K, Murakami A. Hamiltonian Monte Carlo for simultaneous interface and spatial field detection (HMCSISFD) and its application to a piping zone interface detection problem. International Journal for Numerical and Analytical Methods in Geomechanics. 2021;**45**:2602-2626. DOI: 10.1002/NAG.3279

[9] Koch MC, Fujisawa K, Murakami A. Adjoint Hamiltonian Monte Carlo algorithm for the estimation of elastic modulus through the inversion of elastic wave propagation data. International Journal for Numerical Methods in Engineering. 2020;**121**:1037-1067. DOI: 10.1002/nme.6256

[10] Koch MC, Fujisawa K, Murakami A. Novel parameter update for a gradient based MCMC method for solid-void interface detection through elastodynamic inversion. Probabilistic Engineering Mechanics. 2020;**62**: 103097. DOI: 10.1016/j.probengmech.2020.103097

[11] Loève M. Probability Theory II. New York: Springer-Verlag; 1978

[12] Ghanem RG, Spanos PD. Stochastic Finite Elements: A Spectral Approach. Springer New York: New York, NY; 1991. DOI: 10.1007/978-1-4612-3094-6

[13] Sudret B, Der Kiureghian A. Stochastic Finite Element Methods and Reliability: A State-of-the-Art Report. UCB/SEMM-2000/08. Berkeley: University of California; 2000

[14] Betz W, Papaioannou I, Straub D. Numerical methods for the discretization of random fields by means of the Karhunen–Loève expansion. Computer Methods in Applied Mechanics and Engineering. 2014;**271**: 109-129. DOI: 10.1016/J.CMA.2013.12.010

[15] Marzouk YM, Najm HN. Dimensionality reduction and

polynomial chaos acceleration of Bayesian inference in inverse problems. Journal of Computational Physics. 2009; **228**:1862-1902. DOI: 10.1016/j.jcp.2008.11.024

[16] Leimkuhler B, Reich S. Simulating Hamiltonian Dynamics. Cambridge: Cambridge University Press; 2005. DOI: 10.1017/CBO9780511614118

[17] Betancourt M. A conceptual introduction to hamiltonian Monte Carlo. ArXiv Preprint. 2017

[18] Andrieu C, De Freitas N, Doucet A, Jordan MI. An introduction to MCMC for machine learning. Machine Learning. 2003:5-43. DOI: 10.1023/A:1020281327116

[19] Stein K, Tezduyar T, Benney R. Mesh moving techniques for fluid-structure interactions with large displacements. Journal of Applied Mechanics. 2003;**70**:58-63. DOI: 10.1115/1.1530635

[20] Koch MC, Osugi M, Fujisawa K, Murakami A. Investigation of reference parameter invariance and application of HMCSISFD for identification of an eroded seepage zone. Proc. 13th Int. Conf. Struct. Saf. Reliab. 2021–2022, Shanghai. 2022

[21] Magnus JR. On Differentiating Eigenvalues and Eigenvectors. Econometric Theory. 1985;**1**:179-191. DOI: 10.1017/S0266466600011129

Chapter 5

Solving and Algorithm for Least-Norm General Solution to Constrained Sylvester Matrix Equation

Abdur Rehman and Ivan I. Kyrchei

Abstract

Keeping in view that a lot of physical systems with inverse problems can be written by matrix equations, the least-norm of the solution to a general Sylvester matrix equation with restrictions $A_1X_1 = C_1, X_1B_1 = C_2,\ \ A_2X_2 = C_3, X_2B_2 = C_4,\ \ A_3X_1B_3 + A_4X_2B_4 = C_c,$ is researched in this chapter. A novel expression of the general solution to this system is established and necessary and sufficient conditions for its existence are constituted. The novelty of the proposed results is not only obtaining a formal representation of the solution in terms of generalized inverses but the construction of an algorithm to find its explicit expression as well. To conduct an algorithm and numerical example, it is used the determinantal representations of the Moore–Penrose inverse previously obtained by one of the authors.

Keywords: linear matrix equation, generalized Sylvester matrix equation, Moore-Penrose inverse

1. Introduction

Standardly, we state $\mathbb{C}$ and $\mathbb{R}$, respectively, for the complex and real numbers. Let $\mathbb{C}^{m\times n}$ denote the set of all $m \times n$ matrices over $\mathbb{C}$, and $\mathbb{C}_r^{m\times n}$ stay for a subset of $m \times n$ complex matrices with rank r. The rank of A is denoted by both symbols $r(A)$ and $rankA$. The (complex) conjugate transpose matrix of $A \in \mathbb{C}^{m\times n}$ is written by A^* and a matrix $A \in \mathbb{C}^{n\times n}$ is said to be Hermitian if $A^* = A$. An identity matrix with feasible shape is denoted by I.

Definition 1.1. The Moore–Penrose (MP-) inverse of $A \in \mathbb{C}^{m\times n}$, denoted by $A^\dagger$, is defined to be the unique solution X to the following four Penrose equations

$$AXA = A, \tag{1}$$

$$XAX = X, \tag{2}$$

$$(AX)^* = AX, \tag{3}$$

$$(XA)^* = XA. \tag{4}$$

Matrices satisfying the eqs. (1) and (2) are known as reflexive inverses, denoted by A^{+}.

In addition, $L_A = I - A^{\dagger}A$ and $R_A = I - AA^{\dagger}$ represent a pair of orthogonal projectors onto the kernels of A and A^*, respectively.

Mathematical models of physical systems with inverse problems especially those has a finite number of model parameters can be written by matrix equations. In particular, the Sylvester-type matrix equations have far-reaching applications in singular system control [1], system design [2], robust control [3], feedback [4], perturbation theory [5], linear descriptor systems [6], neural networks [7] and theory of orbits [8], etc.

Some recent work on generalized Sylvester matrix equations and their systems can be observed in [9–21]. In 2014, Bao [22] examined the least-norm and extremal ranks of the least square solution to the quaternion matrix equations

$$A_1X = C_1, XB_1 = C_2, \quad A_3XB_3 = C_c. \tag{5}$$

Wang et al. [23] examined the expression of the general solution to the system

$$A_1X_1 = C_1, \quad A_2X_2 = C_3, A_3X_1 \quad B_3 + A_4X_2B_4 = C_c, \tag{6}$$

and as an application, the P-symmetric and P-skew-symmetric solution to

$$A_aX = C_a, A_bXB_b = C_b.$$

has been established. Li et al. [24] established a novel expression of the general solution of the system (6) and they computed the least-norm of general solution to (6). In 2009, Wang et al. [25] constituted the expression of the general solution to

$$\begin{aligned} A_1 \quad & X_1 = C_1, X_1B_1 = C_2, \\ A_2 \quad & X_2 = C_3, X_2B_2 = C_4, \\ A_3 \quad & X_1B_3 + A_4X_2B_4 = C_c, \end{aligned} \tag{7}$$

and as an application, they explored the (P, Q)-symmetric solution to the system

$$A_aX = C_a, XB_b = C_b, A_cXB_c = C_c.$$

Some latest findings on the least-norm of matrix equations and (P, Q)-symmetric matrices can be consulted in [26–30]. Furthermore, our main system (7) is a special case of the following system

$$\begin{aligned} A_1X_1 &= C_1, X_2B_1 = D_1, \\ A_2X_3 &= C_2, X_3B_2 = D_2, \\ A_3X_4 &= C_3, X_4B_3 = D_3, \\ A_4X_1 + X_2 \quad B_4 + C_4X_3D_4 + C_5X_4D_5 &= C_c, \end{aligned} \tag{8}$$

which has been investigated by Zhang in 2014.

Motivated by the latest interest of least-norm of matrix equations, we construct a novel expression of the general solution to the system (7) and apply this to investigate the least-norm of the general solution to the system (7) in this chapter. Observing that systems (5) and (6) are particular cases of our system (7), solving system (7) will encourage the least-norm to a wide class of problems.

We commence with the following lemmas which have crucial function in the construction of the chief outcomes of the following sections.

Lemma 1.2. *[31]. Let A, B, and C be given matrices over $\mathbb{C}$ with agreeable dimensions.* Then.

1. $r(A) + r(R_A B) = r(B) + r(R_B A) = r[A \quad B]$.

2. $r(A) + r(CL_A) = r(C) + r(AL_C) = r\begin{bmatrix} A \\ C \end{bmatrix}$.

3. $r(B) + r(C) + r(R_B A L_C) = r\begin{bmatrix} A & B \\ C & 0 \end{bmatrix}$.

Lemma 1.3. *[32]. Let A, B, and C be known matrices over $\mathbb{C}$ with right sizes. Then*

1. $A^\dagger = (A^* A)^\dagger A^* = A^* (AA^*)^\dagger$.

2. $L_A = L_A^2 = L_A^*, R_A = R_A^2 = R_A^*$.

3. $L_A (BL_A)^\dagger - (BL_A)^\dagger, (R_A C)^\dagger R_A = (R_A C)^\dagger$.

Lemma 1.4. *[33]. Let Φ, Ω be matrices over $\mathbb{C}$ and*

$$\Phi = \begin{bmatrix} \Phi_1 \\ \Phi_2 \end{bmatrix}, \quad \Omega = \begin{bmatrix} \Omega_1 & \Omega_2 \end{bmatrix}, \quad F = \Phi_2 L_{\Phi_1}, \quad T = R_{\Omega_1} \Omega_2.$$

Then

$$L_\Phi = L_{\Phi_1} L_F, \quad L_\Omega = \begin{bmatrix} L_{\Omega_1} & -\Omega_1^\dagger \Omega_2 L_T \\ 0 & L_T \end{bmatrix},$$
$$R_\Omega = R_T R_{\Omega_1}, \quad R_\Phi = \begin{bmatrix} R_{\Phi_1} & 0 \\ -R_F \Phi_2 \Phi_1^\dagger & R_F \end{bmatrix},$$

where Φ_1^+, Ω_1^+ are any fixed reflexive inverses, L_{Φ_1} and R_{Ω_1} stand for the projectors $L_{\Phi_1} = I - \Phi_1^+ \Phi_1$, $R_{\Omega_1} = I - \Omega_1 \Omega_1^+$ induced by Φ_1, Ω_1, respectively.

Remark 1.5. Since the Moore-Penrose inverse is a reflexive inverse, this lemma can be used for the MP-inverse without any changes. It has taken place in ([32], Lemma 2.4).

Lemma 1.6. *[34]. Suppose that*

$$B_1 X C_1 + B_2 Y C_2 = A \tag{9}$$

is consistent linear matrix equation. Then.

1. *The general solution of the homogeneous equation*

$$B_1 X C_1 + B_2 Y C_2 = 0,$$

can be expressed by

$$X = X_1X_2 + X_3, \quad Y = Y_1Y_2 + Y_3,$$

where $X_1 - X_3$ *and* $Y_1 - Y_3$ *are general solution to the system*

$$B_1X_1 = -B_2Y_1, \ X_2C_1 = Y_2C_2, \ B_1X_3C_1 = 0, \ B_2Y_3C_2 = 0.$$

By computing the value of unknowns in above and using them in X and Y, *we have*

$$X = S_1L_GUR_HT_1 + L_{B_1}V_1 + V_2R_{C_1},$$
$$Y = S_2L_GUR_HT_2 + L_{B_2}W_1 + W_2R_{C_2},$$

where $S_1 = \begin{bmatrix} I_p, & 0 \end{bmatrix}, S_2 = [0, \ I_s], T_1 = \begin{bmatrix} I_q \\ 0 \end{bmatrix}, \ T_2 = \begin{bmatrix} 0 \\ I_t \end{bmatrix}, G = [B_1, \ B_2]$, *and*

$H = \begin{bmatrix} C_1 \\ -C_2 \end{bmatrix}$; *the matrices* U, V_1, V_2, W_1 *and* W_2 *are free to vary over* $\mathbb{C}$.

2. *Assume that* Eq. (9) *is solvable, then its general solution can be expressed as*

$$X = X_0 + X_1X_2 + X_3, \quad Y = Y_0 + Y_1Y_2 + Y_3,$$

where X_0 *and* Y_0 *are any pair of particular solutions to (9).*

It can also be written as

$$X = X_0 + S_1L_GUR_HT_1 + L_{B_1}V_1 + V_2R_{C_1},$$
$$Y = Y_0 + S_2L_GUR_HT_2 + L_{B_2}W_1 + W_2R_{C_2}.$$

Lemma 1.7. *[35]. Let* A_1, B_1, C_1, C_2 *be given matrices over* $\mathbb{C}$ *with agreeable sizes and* X_1 *to be determined. Then the system*

$$A_1X_1 = C_1, X_1B_1 = C_2, \tag{10}$$

is consistent if and only if

$$R_{A_1}C_1 = 0, \ C_2L_{B_1} = 0, \ A_1C_2 = C_1B_1. \tag{11}$$

Under these conditions, the general solution to (10) *can be established as*

$$X_1 = A_1^{\dagger}C_1 + L_{A_1}C_2B_1^{\dagger} + L_{A_1}U_1R_{B_1},$$

where U_1 *is a free matrix over* $\mathbb{C}$ *with accordant dimension.*

Lemma 1.8. *[36]. Let A, B, and C be known matrices over* $\mathbb{C}$ *with agreeable dimensions, and X be unknown. Then the matrix equation*

$$AXB = C \tag{12}$$

is consistent if and only if $AA^{\dagger}CB^{\dagger}B = C$. *In this case, its general solution can be expressed as*

$$X = A^{\dagger}CB^{\dagger} + L_A V + WR_B, \tag{13}$$

where V, W are arbitrary matrices over $\mathbb{C}$ with appropriate dimensions.

In [37], it is proved that (13) is the least squares solution to (12), and its minimum norm least squares solution is $X_{LS} = A^{\dagger}CB^{\dagger}$.

Lemma 1.9. *[25]. Let A_i, B_i, C_i, $(i = 1, \ldots, 4)$, and C_c be given matrices over $\mathbb{C}$ with agreeable dimensions, and X_1, X_2 to be determined. Denote*

$$\begin{aligned}
&A = A_3 L_{A_1}, B = R_{B_1} B_3, C = A_4 L_{A_2}, D = R_{B_2} B_4,\\
&N = DL_B, M = R_A C, S = CL_M,\\
&E = C_c - A_3 A_1^{\dagger} C_1 B_3 - AC_2 B_1^{\dagger} B_3 - A_4 A_2^{\dagger} C_3 B_4 - CC_4 B_2^{\dagger} B_4.
\end{aligned}$$

Then the following conditions are tantamount:

1. *System* (7) *is resolvable.*

2. *The conditions in* (11) *are met and*

$$\begin{aligned}
&R_{A_2} C_3 = 0, \quad C_4 L_{B_2} = 0, \quad A_2 C_4 = C_3 B_2,\\
&R_M R_A E = 0, R_A E L_D = 0, E L_B L_N = 0, R_C E L_B = 0.
\end{aligned} \tag{14}$$

3. *The equalities in* (11) and (14) *are satisfied and*

$$MM^{\dagger} R_A D^{\dagger} D = R_A E, \quad CC^{\dagger} E L_B N^{\dagger} N = E L_B.$$

In these conditions, the general solution to the system (7) *can be written as*

$$\begin{aligned}
X_1 = {} & A_1^{\dagger} C_1 + L_{A_1} C_2 B_1^{\dagger} + L_{A_1} A^{\dagger} E B^{\dagger} R_{B_1} - L_{A_1} A^{\dagger} C M^{\dagger} E B^{\dagger} R_{B_1} -\\
& - L_{A_1} A^{\dagger} S C^{\dagger} E N^{\dagger} D B^{\dagger} R_{B_1} - L_{A_1} A^{\dagger} S V_1 R_N D B^{\dagger} R_{B_1} +\\
& + L_{A_1} (L_A U_1 + Z_1 R_B) R_{B_1},
\end{aligned} \tag{15}$$

$$\begin{aligned}
X_2 = {} & A_2^{\dagger} C_3 + L_{A_2} C_4 B_2^{\dagger} + L_{A_2} M^{\dagger} R_A E D^{\dagger} R_{B_2} + L_{A_2} L_{M_b} S^{\dagger} S C^{\dagger} E N^{\dagger} R_{B_2}\\
& + L_{A_2} L_M (V_1 - S^{\dagger} S V_1 N N^{\dagger}) R_{B_2} + L_{A_2} W_1 R_D R_{B_2},
\end{aligned} \tag{16}$$

where U_1, V_1, W_1 and Z_1 are free matrices over $\mathbb{C}$ with agreeable dimensions.

Since the general solutions of considered systems are expressed in terms of generalized inverses, another goal of the paper is to give determinantal representations of the least-norm of the general solution to the system (7) based on determinantal representations of generalized inverses.

Due to the important role of generalized inverses in many application fields, considerable effort has been exerted toward the numerical algorithms for fast and accurate calculation of matrix generalized inverse. In general, most existing methods for their obtaining are iterative algorithms for approximating generalized inverses of complex matrices (some recent papers, see, e.g. [38–40]). There are only several direct methods for finding MP-inverse for an arbitrary complex matrix $A \in \mathbb{C}^{m \times n}$. The most famous is method based on singular value decomposition (SVD), i.e. if $A = U\Sigma V^*$, then $A^{\dagger} = V\Sigma^{\dagger} U^*$. The computational cost of this method is dominated by the cost of computing the SVD, which is several times higher than matrix–matrix multiplication. Another approach is constructing determinantal representations of the MP-inverse $A^{\dagger}$. A well-known determinantal representation of an ordinary inverse is the adjugate matrix with the cofactors in entries. It has an important theoretical significance and brings forth Cramer's rule for systems of linear equations. The same is desirable to have for the generalized inverses. Due to looking for their more applicable explicit expressions, there are various determinantal representations of generalized inverses (for the MP-inverse, see, e.g. [41, 42]). Because of the complexity of the previously obtained expressions of determinantal representations of the MP-inverse, they have little applicability.

In this chapter, we will use the determinantal representations of the MP-inverse recently obtained in [43].

Lemma 1.10. *[43, Theorem 2.2] If $A \in \mathbb{C}^{m\times n}$ with rank$A = r$, then the Moore-Penrose inverse $A^{\dagger} = \left(a_{ij}^{\dagger}\right) \in \mathbb{C}^{n\times m}$ possess the following determinantal representations*

$$a_{ij}^{\dagger} = \frac{\sum_{\beta \in J_{r,n}\{i\}} \left|(A^*A)_{.i}\left(a_{.j}^*\right)\right|_{\beta}^{\beta}}{\sum_{\beta \in J_{r,n}} |A^*A|_{\beta}^{\beta}} = \frac{\sum_{\alpha \in I_{r,m}\{j\}} \left|(AA^*)_{j.}\left(a_{i.}^*\right)\right|_{\alpha}^{\alpha}}{\sum_{\alpha \in I_{r,m}} |AA^*|_{\alpha}^{\alpha}}. \tag{17}$$

Here $|A|_{\alpha}^{\alpha}$ denote a principal minor of A whose rows and columns are indexed by $\alpha := \{\alpha_1, \ldots, \alpha_k\} \subseteq \{1, \ldots, m\}$,

$$L_{k,m} := \{\alpha : 1 \le \alpha_1 < \cdots < \alpha_k \le m\}, \text{and } I_{r,m}\{i\} := \{\alpha : \alpha \in L_{r,m}, i \in \alpha\}.$$

Also, $a_{.j}^*$ and $a_{i.}^*$ denote the jth column and the ith row of A^*, and $A_{i.}(b)$ and $A_{.j}(c)$ stand for the matrices obtained from A by replacing its ith row with the row vector $b \in \mathbb{C}^{1\times n}$ and its jth column with the column vector $c \in \mathbb{C}^m$, respectively.

The formulas (17) give very simple and elegant determinantal representations of the MP-inverse. So, for any $A \in \mathbb{C}_r^{m\times n}$, we have sum of all principal minors of r order of the matrices A^*A or AA^* in denominators and sum of principal minors of r order of the matrices $(A^*A)_{.i}\left(a_{.j}^*\right)$ or $(AA^*)_{j.}\left(a_{i.}^*\right)$ that contain the ith column or the jth row, respectively, in numerators into (17).

Note that for an arbitrary full-rank matrix A, Lemma 1.10 gives a new way of finding an inverse matrix.

Corollary 1.11. *If $A \in \mathbb{C}^{m\times n}$ with rank$A = \min\{m, n\}$, then the inverse $A^{-1} = \left(a_{ij}^{-1}\right) \in \mathbb{C}^{n\times m}$ possess the following determinantal representations:*

$$a_{ij}^{-1} = \begin{cases} \dfrac{\left|(A^*A)_{.i}\left(a_{.j}^*\right)\right|}{|A^*A|} & \text{if } \operatorname{rank}A = n, \\[2ex] \dfrac{\left|(AA^*)_{j.}\left(a_{i.}^*\right)\right|}{|AA^*|} & \text{if } \operatorname{rank}A = m. \end{cases}$$

These new determinantal representations of the Moore-Penrose inverse have been obtained by the developed novel limit-rank method in the case of quaternion matrices [44] as well. This method was successfully applied for constructing determinantal representations of other generalized inverses in both cases for complex and quaternion matrices (see e.g. [45–47]). It also yields Cramer's rules of various matrix equations [48–54].

The remainder of our chapter is directed as follows. In Section 2, we provide a new expression of the general solution to our system (7) and discuss its least-norm. The algorithm and numerical example of finding the anti-Hermitian solution to (7) are presented in Section 3. (7). Finally, in Section 4, the conclusions are drawn.

2. A new expression of the general solution to the system

Now we demonstrate the principal theorem of this section (7).

Theorem 2.1. *Assume that* $S_1 = [I_{p_1}\ 0], S_2 = [0\ I_{p_2}]$, $T_1 = \begin{bmatrix} I_{q_1} \\ 0 \end{bmatrix}$, $T_2 = \begin{bmatrix} I_{q_2} \\ 0 \end{bmatrix}$, $G = [A\ C], H = \begin{bmatrix} B \\ -D \end{bmatrix}$, $H_1 = L_{A_1}L_A, H_2 = L_{A_1}S_1L_G, H_3 = R_HT_1R_{B_1}, H_4 = L_{A_2}L_C, H_5 = L_{A_2}S_2L_G, H_6 = R_HT_2R_{B_2}$ *and the system* (7) *is solvable, then the general solution to our system can be formed as*

$$\begin{aligned} X_1 = {} & A_1^\dagger C_1 + L_{A_1}C_2B_1^\dagger + L_{A_1}A^\dagger EB^\dagger R_{B_1} - L_{A_1}A^\dagger CM^\dagger EB^\dagger R_{B_1} - \\ & -L_{A_1}A^\dagger SC^\dagger EN^\dagger DB^\dagger R_{B_1} + H_1V_1R_{B_1} + H_2UH_3 + L_{A_1}V_2R_BR_{B_1}, \end{aligned} \tag{18}$$

$$\begin{aligned} X_2 = {} & A_2^\dagger C_3 + L_{A_2}C_4B_2^\dagger + L_{A_2}M^\dagger R_AED^\dagger R_{B_2} + L_{A_2}L_MS^\dagger SC^\dagger EN^\dagger R_{B_2} + \\ & +H_4W_1R_{B_2} + H_5UH_6 + L_{A_2}W_2R_DR_{B_2}, \end{aligned} \tag{19}$$

where $U, V_1, V_2, W_1,$ *and* W_2 *are free matrices over* $\mathbb{C}$ *with allowable dimensions.*

Proof. Our proof contains three parts. At the first step, we show that the matrices X_1 and X_2 have the forms of

$$X_1 = \phi_0 + H_1V_1R_{B_1} + L_{A_1}V_2R_BR_{B_1} + H_2UH_3, \tag{20}$$

$$X_2 = \psi_0 + H_4W_1R_{B_2} + L_{A_2}W_2R_DR_{B_2} + H_5UH_6, \tag{21}$$

where ϕ_0 and ψ_0 are any pair of particular solution to the system (7), V_1, V_2, W_1, W_2, and U are free matrices of able shapes over $\mathbb{C}$, are solutions to the system (7). In the second step, we display that any couple of solutions μ_0 and ν_0 to the system (7) can be established as (20) and (21), respectively. In the end, we confirm that

$$\begin{aligned} \mu &= A_1^\dagger C_1 + L_{A_1}C_2B_1^\dagger + A^\dagger EB^\dagger - A^\dagger CM^\dagger EB^\dagger - A^\dagger SC^\dagger EN^\dagger DB^\dagger, \\ \nu &= A_2^\dagger C_3 + L_{A_2}C_4B_2^\dagger + L_{A_2}M^\dagger R_AED^\dagger + L_{A_2}L_MS^\dagger SC^\dagger EN^\dagger R_{B_2} \end{aligned}$$

are a couple of particular solutions to the system (7).

Now we prove that a couple of matrices X_1 and X_2 having the shape of (20) and (21), respectively, are solutions to the system (7). Observe that

$$\begin{aligned} A_1^\dagger C_1B_1 + L_{A_1}C_2B_1^\dagger B_1 &= A_1^\dagger A_1C_2 + L_{A_1}C_2 = C_2, \\ A_2^\dagger C_3B_2 + L_{A_2}C_4B_2^\dagger B_2 &= A_2^\dagger A_2C_4 + L_{A_2}C_4 = C_4. \end{aligned}$$

It is evident that X_1 having the form (20) is a solution of $A_1X_1 = C_1$, and $X_1B_1 = C_2$ and X_2 having the form (21) is a solution to $A_2X_2 = C_3, X_2B_2 = C_4$. Now we are left to show that $A_3X_1B_3 + A_4X_2B_4 = C_c$ is satisfied by X_1 and X_2 given in (20) and (21). By Lemma 1.4, we have

$$\begin{aligned} AS_1L_G &= A\begin{bmatrix} I_{p_1} & 0 \end{bmatrix}\begin{bmatrix} L_A & -A^{\dagger}CL_M \\ 0 & L_M \end{bmatrix} = A\begin{bmatrix} L_A & -A^{\dagger}CL_M \end{bmatrix} \\ &= \begin{bmatrix} 0 & -AA^{\dagger}CL_M \end{bmatrix} = \begin{bmatrix} 0 & -(C-M)L_M \end{bmatrix} = \begin{bmatrix} 0 & -CL_M \end{bmatrix} \\ &= -\begin{bmatrix} 0 & S \end{bmatrix} = -CS_2L_G, \end{aligned} \tag{22}$$

and

$$\begin{aligned} R_HT_1B &= \begin{bmatrix} R_B & 0 \\ R_NDB^{\dagger} & R_N \end{bmatrix}\begin{bmatrix} I_{q_1} \\ 0 \end{bmatrix}B = \begin{bmatrix} R_B \\ R_NDB^{\dagger} \end{bmatrix}B \\ &= \begin{bmatrix} 0 \\ R_NDB^{\dagger}B \end{bmatrix} = \begin{bmatrix} 0 \\ R_ND(I-L_B) \end{bmatrix} = \begin{bmatrix} 0 \\ R_ND \end{bmatrix} \\ &= R_HT_2D. \end{aligned} \tag{23}$$

Observe that $AL_A = 0$ and by using (22) and (23), we arrive that

$$A_3X_1B_3 + A_4X_2B_4 = C_c.$$

Conversely, assume that μ_0 and ν_0 are any couple of solutions to our system (7). By Lemma 1.7, we have

$$\begin{aligned} &A_1A_1^{\dagger}C_1 = C_1, C_2B_1^{\dagger}B_1 = C_2, A_2A_2^{\dagger}C_3 = C_3, \\ &C_4B_2^{\dagger}B_2 = C_4, A_1C_2 = C_1B_1, A_2C_4 = C_3B_2. \end{aligned}$$

Observe that

$$\begin{aligned} L_{A_1}\mu_0R_{B_1} &= (I - A_1^{\dagger}A_1)\mu_0(I - B_1B_1^{\dagger}) \\ &= \mu_0 - \mu_0B_1B_1^{\dagger} - A_1^{\dagger}A_1\mu_0 + A_1^{\dagger}A_1\mu_0B_1B_1^{\dagger} \\ &= \mu_0 - C_2B_1^{\dagger} - A_1^{\dagger}C_1 + A_1^{\dagger}A_1C_2B_1^{\dagger} \\ &= \mu_0 - L_{A_1}C_2B_1^{\dagger} - A_1^{\dagger}C_1 \end{aligned}$$

produces

$$\mu_0 = L_{A_1}C_2B_1^{\dagger} + A_1^{\dagger}C_1 + L_{A_1}\mu_0R_{B_1}. \tag{24}$$

On the same lines, we can get

$$\nu_0 = L_{A_2}C_4B_2^{\dagger} + A_2^{\dagger}C_3 + L_{A_2}\nu_0R_{B_2}. \tag{25}$$

It is manifest that μ_0 and ν_0 defined in (24)–(25) are also solution pair of

$$AX_1B + CX_2D = E. \tag{26}$$

Since

$$\begin{aligned} AX_1B + CX_2D &= A_3L_{A_1}\mu_0R_{B_1}B_3 + A_4L_{A_2}\nu_0R_{B_2}B_4 \\ &= A_3\left(\mu_0 - L_{A_1}C_2B_1^{\dagger} - A_1^{\dagger}C_1\right)B_3 + A_4\left(\nu_0 - L_{A_2}C_4B_2^{\dagger} - A_2^{\dagger}C_3\right)B_4 \\ &= A_3\mu_0B_3 - A_3L_{A_1}C_2B_1^{\dagger}B_3 - A_1^{\dagger}C_1B_3 + A_4\nu_0B_4 \\ &\quad -A_4L_{A_2}C_4B_2^{\dagger}B_4 - A_4A_2^{\dagger}C_3B_4 \\ &= A_3\mu_0B_3 + A_4\nu_0B_4 - AC_2B_1^{\dagger}B_3 - A_1^{\dagger}C_1B_3 - CC_4B_2^{\dagger}B_4 - A_4A_2^{\dagger}C_3B_4 \\ &= C_c - AC_2B_1^{\dagger}B_3 - A_1^{\dagger}C_1B_3 - CC_4B_2^{\dagger}B_4 - A_4A_2^{\dagger}C_3B_4 \\ &= E. \end{aligned}$$

Hence by Lemma 1.6, μ_0 and ν_0 can be written as

$$\mu_0 = X_{01} + S_1L_GUR_HT_1 + L_AV_1 + V_2R_B, \tag{27}$$

$$\nu_0 = X_{02} + S_2L_GUR_HT_2 + L_CW_1 + W_2R_D, \tag{28}$$

where X_{01} and X_{02} are a couple of special solutions to (26) and U, V_1, V_2, W_1 and W_2 are free matrices with agreeable dimensions. Using (27) and (28) in (24) and (25), respectively, we get

$$\begin{aligned} \mu_0 &= X_{10} + H_2UH_3 + H_1V_1R_{B_1} + L_{A_1}V_2R_BR_{B_1}, \\ \nu_0 &= X_{20} + H_5UH_6 + H_4W_1R_{B_2} + L_{A_2}W_2R_DR_{B_2}, \end{aligned}$$

where $X_{10} = A_1^{\dagger}C_1 + L_{A_1}C_2B_1^{\dagger} + L_{A_1}X_{01}R_{B_1}$ and $X_{20} = A_2^{\dagger}C_3 + L_{A_2}C_4B_2^{\dagger} + L_{A_2}X_{02}R_{B_2}$. It is evident that X_{10} and X_{20} are a couple of solutions to the system (7). It is clear that μ_0 and ν_0 can be represented by (20) and (21), respectively. Lastly, by putting U_1, V_1, W_1, and Z_1 equal to zero in (15) and (16), we conclude that μ and ν are special solutions to the system (7). Hence the expressions (18) and (19) represent the general solution to the system (7) and the theorem is completed.

Remark 2.2. Due to Lemma 1.3 and taking into account $L_{A_2}L_M = L_ML_{A_2}$, we have the following simplification of the solution pair to the system (7) that is identical for (15)–(16) and (18)–(19) when $U, U_1, V_1, V_2, Z_1, W_1$, and W_2 disappear,

$$\begin{aligned} X_1 &= A_1^{\dagger}C_1 + L_{A_1}C_2B_1^{\dagger} + A^{\dagger}EB^{\dagger} - A^{\dagger}A_4M^{\dagger}EB^{\dagger} - A^{\dagger}SC^{\dagger}EN^{\dagger}B_4B^{\dagger}, \\ X_2 &= A_2^{\dagger}C_3 + L_{A_2}C_4B_2^{\dagger} + M^{\dagger}ED^{\dagger} + S^{\dagger}SC^{\dagger}EN^{\dagger}. \end{aligned}$$

Comment 2.3. We have established a novel expression of the general solution to the system (7) in Theorem 2.1 which is different from one created in [25]. With the help of this novel expression, we can explore the least-norm of the general solution which can not be studied with the help of the expression given in [25], which is one of the advantage of our new expression.

Now we discuss some special cases of our system.

If B_1, B_2, C_2 and C_4 disappear in Theorem 2.1, then we gain the following conclusion.

Corollary 2.4. *Denote* $S_1 = \begin{bmatrix} I_{p_1} & 0 \end{bmatrix}, S_2 = \begin{bmatrix} 0 & I_{p_2} \end{bmatrix}, T_1 = \begin{bmatrix} I_{q_1} \\ 0 \end{bmatrix}, T_2 = \begin{bmatrix} I_{q_2} \\ 0 \end{bmatrix},$ $G = \begin{bmatrix} A & C \end{bmatrix}, H = \begin{bmatrix} B_3 \\ -B_4 \end{bmatrix}, H_1 = L_{A_1}L_A, H_2 = L_{A_1}S_1L_G, H_3 = R_HT_1, H_4 = L_{A_2}L_C, H_5 =$

$L_{A_2}S_2L_G, H_6 = R_HT_2$ *and the system* (6) *is solvable, then the general solution to* (6) *can be formed as*

$$\begin{aligned} X_1 &= A_1^\dagger C_1 + A^\dagger EB_3^\dagger - A^\dagger A_4M^\dagger EB_3^\dagger - A^\dagger SC^\dagger EN^\dagger B_4B_3^\dagger - H_1Y_1 + \\ &\quad + H_2VH_3 + L_{A_1}Y_2R_{B_3}, \\ X_2 &= A_2^\dagger C_3 + M^\dagger EB_4^\dagger + S^\dagger SC^\dagger EN^\dagger + H_4Z_1 + H_5VH_6 + L_{A_2}Z_2R_{B_4}, \end{aligned}$$

where A, C, N, M, S *are the same as in Lemma 1.6,* $E = C_c - A_3A_1^\dagger C_1B_3 - A_4A_2^\dagger C_3B_4$, $V, Y_1, Y_2, Z_1,$ and Z_2 *are free matrices over* $\mathbb{C}$ *obeying agreeable dimensions.*

Comment 2.5. The above consequence is a chief result of [32].

If A_2, B_2, C_3, A_4, B_4 and C_4 vanish in our system (7), then we get the following outcome.

Corollary 2.6. *Suppose that* $A_1, B_1, C_1, C_2, A_3, B_3$ *and* C_c *are given. Then the general solution to system* (5) *is established by*

$$\begin{aligned} X_1 = {} & A_1^\dagger C_1 + L_{A_1}C_2B_1^\dagger + (A_3L_{A_1})^\dagger \left[C_c - A_3A_1^\dagger C_1B_3 - A_3L_{A_1}C_2B_1^\dagger B_3\right](R_{B_1}B_3)^\dagger + \\ & + L_{A_1}L_{A_3L_{A_1}}W_1R_{B_1} + L_{A_1}W_2R_{R_{B_1}B_3}R_{B_1}, \end{aligned}$$

where W_1 and W_2 *are arbitrary matrices over* $\mathbb{C}$ *with appropriate sizes.*

We experience the least-norm to the system (7) in this section. By the definition and [55], we can get the following result easily.

Lemma 2.7. *Let* $A \in \mathbb{C}^{m\times n}, B \in \mathbb{C}^{n\times m}$. *Then we have.*

(1) $\|A + B\|^2 = \|A\|^2 + \|B\|^2 + 2\,Re\left[\mathrm{tr}(B^*A)\right]$.

(2) $Re\left[\mathrm{tr}(AB)\right] = Re\left[\mathrm{tr}(BA)\right]$.

Theorem 2.8. *Assume that system* (7) *is solvable, then the least-norm of the solution pair* X_1 *and* X_2 *to system* (7) *can be extracted as follows:*

$$\begin{aligned} \|X_1\|_{min} &= A_1^\dagger C_1 + L_{A_1}C_2B_1^\dagger + A^\dagger EB^\dagger - A^\dagger A_4M^\dagger EB^\dagger - \\ &\quad - A^\dagger SC^\dagger EN^\dagger B_4B^\dagger, \end{aligned} \tag{29}$$

$$\|X_2\|_{min} = A_2^\dagger C_3 + L_{A_2}C_4B_2^\dagger + M^\dagger ED^\dagger + S^\dagger SC^\dagger EN^\dagger. \tag{30}$$

Proof. By Theorem 2.1 and Remark 2.2, the general solution to (7) can be formed as

$$\begin{aligned} X_1 &= A_1^\dagger C_1 + L_{A_1}C_2B_1^\dagger + A^\dagger EB^\dagger - A^\dagger A_4M^\dagger EB^\dagger - A^\dagger SC^\dagger EN^\dagger B_4B^\dagger \\ &\quad - H_1V_1R_{B_1} + H_2UH_3 + L_{A_1}V_2R_BR_{B_1}, \\ X_2 &= A_2^\dagger C_3 + L_{A_2}C_4B_2^\dagger + M^\dagger ED^\dagger + S^\dagger SC^\dagger EN^\dagger \\ &\quad + H_4W_1R_{B_2} + H_5UH_6 + L_{A_2}W_2R_DR_{B_2}, \end{aligned}$$

where $U, V_1, V_2, W_1,$ and W_2 are free matrices over $\mathbb{C}$ having executable dimensions. By Lemma 2.7, the norm of X_1 can be established as

$$\begin{aligned} \|X_1\|^2 = {} & \|A_1^\dagger C_1 + L_{A_1}C_2B_1^\dagger + A^\dagger EB^\dagger - A^\dagger A_4M^\dagger EB^\dagger - \\ & - A^\dagger SC^\dagger EN^\dagger B_4B^\dagger - H_1V_1R_{B_1} + H_2UH_3 + L_{A_1}V_2R_BR_{B_1}\|^2 \\ = {} & \|A_1^\dagger C_1 + L_{A_1}C_2B_1^\dagger + A^\dagger EB^\dagger - A^\dagger A_4M^\dagger EB^\dagger - A^\dagger SC^\dagger EN^\dagger B_4B^\dagger\|^2 \\ & + \|H_1V_1R_{B_1} + H_2UH_3 + L_{A_1}V_2R_BR_{B_1}\|^2 + J, \end{aligned} \tag{31}$$

where

$$J = 2\,Re\left[\mathrm{tr}((H_1V_1R_{B_1} + H_2UH_3 + L_{A_1}V_2R_BR_{B_1})^* \right. \\ \left.(A_1^\dagger C_1 + L_{A_1}C_2B_1^\dagger + A^\dagger EB^\dagger - A^\dagger A_4M^\dagger EB^\dagger - A^\dagger SC^\dagger EN^\dagger B_4B^\dagger))\right]. \tag{32}$$

Now we want to show that $J = 0$. Applying Lemmas 1.3, 1.4 and 2.7, we have

$$\begin{aligned} & Re\left[\mathrm{tr}\left((H_1V_1R_{B_1})^*(A_1^\dagger C_1 + L_{A_1}C_2B_1^\dagger + A^\dagger EB^\dagger - A^\dagger A_4M^\dagger EB^\dagger\right.\right. \\ & \left.\left.-A^\dagger SC^\dagger EN^\dagger B_4B^\dagger))\right] = Re\left[\mathrm{tr}\left(R_{B_1}V_1^*H_1^*(A_1^\dagger C_1 + L_{A_1}C_2B_1^\dagger + A^\dagger EB^\dagger\right.\right. \\ & \left.\left.-A^\dagger A_4M^\dagger EB^\dagger - A^\dagger SC^\dagger EN^\dagger B_4B^\dagger))\right] = Re\left[\mathrm{tr}\left(R_{B_1}V_1^*L_AL_{A_1}(A_1^\dagger C_1\right.\right. \\ & \left.\left.+L_{A_1}C_2B_1^\dagger + A^\dagger EB^\dagger - A^\dagger A_4M^\dagger EB^\dagger - A^\dagger SC^\dagger EN^\dagger B_4B^\dagger))\right] \\ & = Re\left[\mathrm{tr}\left(R_{B_1}V_1^*L_AL_{A_1}(L_{A_1}C_2B_1^\dagger)\right)\right] \\ & = Re\left[\mathrm{tr}\left(V_1^*L_AL_{A_1}(L_{A_1}C_2B_1^\dagger)R_{B_1}\right)\right] = 0, \end{aligned} \tag{33}$$

$$\begin{aligned} & Re\left[\mathrm{tr}\left((L_{A_1}V_2R_BR_{B_1})^*(A_1^\dagger C_1 + L_{A_1}C_2B_1^\dagger + A^\dagger EB^\dagger - A^\dagger A_4M^\dagger EB^\dagger\right.\right. \\ & \left.\left.-A^\dagger SC^\dagger EN^\dagger B_4B^\dagger))\right] = Re\left[\mathrm{tr}(R_{B_1}R_BV_2^*L_{A_1}^*(A_1^\dagger C_1 + L_{A_1}C_2B_1^\dagger\right. \\ & \left.+A^\dagger EB^\dagger - A^\dagger A_4M^\dagger EB^\dagger - A^\dagger SC^\dagger EN^\dagger B_4B^\dagger))\right] \\ & = Re\left[\mathrm{tr}\left(R_{B_1}R_BV_2^*L_{A_1}(L_{A_1}C_2B_1^\dagger + A^\dagger EB^\dagger - A^\dagger A_4M^\dagger EB^\dagger\right.\right. \\ & \left.\left.-A^\dagger SC^\dagger EN^\dagger B_4B^\dagger))\right] - Re\left[\mathrm{tr}\left(V_2^*L_{A_1}(L_{A_1}C_2B_1^\dagger\right.\right. \\ & \left.\left.+A^\dagger EB^\dagger - A^\dagger A_4M^\dagger EB^\dagger - A^\dagger SC^\dagger EN^\dagger B_4B^\dagger)R_{B_1}R_B\right)\right] \\ & = Re\left[\mathrm{tr}\left(V_2^*L_{A_1}(A^\dagger EB^\dagger - A^\dagger A_4M^\dagger EB^\dagger - A^\dagger SC^\dagger EN^\dagger B_4B^\dagger)R_B\right)\right] = 0, \end{aligned} \tag{34}$$

$$\begin{aligned} & Re\left[\mathrm{tr}\left((H_2UH_3)^*(A_1^\dagger C_1 + L_{A_1}C_2B_1^\dagger + A^\dagger EB^\dagger - A^\dagger A_4M^\dagger EB^\dagger\right.\right. \\ & \left.\left.-A^\dagger SC^\dagger EN^\dagger B_4B^\dagger))\right] = Re\left[\mathrm{tr}\left(H_3^*U^*H_2^*(A_1^\dagger C_1 + L_{A_1}C_2B_1^\dagger + A^\dagger EB^\dagger\right.\right. \\ & \left.\left.-A^\dagger A_4M^\dagger EB^\dagger - A^\dagger SC^\dagger EN^\dagger B_4B^\dagger))\right] = Re\left[\mathrm{tr}\left(H_3^*U^*L_GS_1^*L_{A_1}(L_{A_1}C_2B_1^\dagger\right.\right. \\ & \left.\left.+A^\dagger EB^\dagger - A^\dagger A_4M^\dagger EB^\dagger - A^\dagger SC^\dagger EN^\dagger B_4B^\dagger))\right] \\ & = Re\left[\mathrm{tr}\left(H_3^*U^*\begin{bmatrix} L_A & -A^\dagger CL_M \\ 0 & L_M \end{bmatrix}\begin{bmatrix} I \\ 0 \end{bmatrix}(L_{A_1}C_2B_1^\dagger + A^\dagger EB^\dagger\right.\right. \\ & \left.\left.-A^\dagger A_4M^\dagger EB^\dagger - A^\dagger SC^\dagger EN^\dagger B_4B^\dagger)\right)\right] = Re\left[\mathrm{tr}\left(H_3^*U^*L_A(A^\dagger EB^\dagger\right.\right. \\ & \left.\left.-A^\dagger A_4M^\dagger EB^\dagger - A^\dagger SC^\dagger EN^\dagger B_4B^\dagger))\right] = Re\left[\mathrm{tr}\left(H_3^*U^*L_AL_{A_1}C_2B_1^\dagger\right)\right] \\ & = Re\left[\mathrm{tr}\left(R_{B_1}T_1^*R_HU^*L_AL_{A_1}C_2B_1^\dagger\right)\right] \end{aligned} \tag{35}$$

By using (33)–(35) in (32) produces $J = 0$. Since X_1 is arbitrary, we get (29) from (31). In the same way, we can prove that (30) hold. □

A special case of our system (7) is given below.

If $B_1, B_2, C_2,$ and C_4 become zero matrices in Theorem 2.8, then again we get the principal result of [20].

Corollary 2.9. *Assume that system* (6) *is solvable, then the least-norm of the solution pair* X_1 *and* X_2 *to system* (6) *can be furnished as*

$$\|X_1\|_{min} = A_1^{\dagger}C_1 + A^{\dagger}EB_3^{\dagger} - A^{\dagger}A_4M^{\dagger}EB_3^{\dagger} - A^{\dagger}SC^{\dagger}EN^{\dagger}B_4B_3^{\dagger},$$
$$\|X_2\|_{min} = A_2^{\dagger}C_3 + M^{\dagger}EB_4^{\dagger} + S^{\dagger}SC^{\dagger}EN^{\dagger}.$$

If A_2, B_2, C_3, A_4, B_4 and C_4 vanish in our system, then we get the next consequence.

Corollary 2.10. *Suppose that* $A_1, B_1, C_1, C_2, A_3, B_3$ *and* C_c *are given. Then the least-norm of the least square solution to system* (5) *is launched by*

$$\|X_1\|_{min} = A_1^{\dagger}C_1 + L_{A_1}C_2B_1^{\dagger} + (A_3L_{A_1})^{\dagger}\left[C_c - A_3A_1^{\dagger}C_1B_3 - A_3L_{A_1}C_2B_1^{\dagger}B_3\right](R_{B_1}B_3)^{\dagger}.$$

Comment 2.11. Corollary 2.10 is the key result of [22].

3. Algorithm with example

In this section, we construct the algorithm for finding the least-norm of the solution to (7) that is inducted by Theorem 2.8.

Algorithm 1.

1. By Lemma 1.10 find the matrices $A_i^{\dagger}$, $B_i^{\dagger}$ for $i = 1, \dots, 4$, and $R_{A_i} = I - A_iA_i^{\dagger}$, $L_{A_i} = I - A_i^{\dagger}A_i$, $R_{B_i} = I - B_iB_i^{\dagger}$, and $L_{B_i} = I - B_i^{\dagger}B_i$ for $i = 1, 2$.

2. By Lemma 1.9 calculate the matrices A, B, C, D, M, S, and E, and by Lemma 1.10 find their MP-inverses and orthogonal projectors when it is needed.

3. Verify the consistence equalities (11) and (14). If these equalities are hold, then we find solutions by the next steps.

4. Finally, by (29) and (30), compute the least-norm of the solution pair X_1 and X_2.

The following example will be considered by using Algorithm 1. Note that our goal is both to confirm correctness of main results from Theorems 2.1 and 2.8, and to demonstrate the technique of applying the determinantal representations of the MP-inverse from Lemma 1.10 by using a not too complicated and understandable example.

Example 1. Given the matrices:

$$A_1 = \begin{bmatrix} 1+i & 1-i & -1+i & -1-i \\ -1+i & 1+i & -1-i & 1-i \\ 2i & 2 & -2 & -2i \end{bmatrix}, B_1 = \begin{bmatrix} 2i & -1 & i+3 \\ -i & 1 & -3-i \\ -1 & i & 1-3i \\ 1 & -i & -1+3i \end{bmatrix}, A_2 = \begin{bmatrix} i & 1 & -1 \\ 1 & -i & i \\ -1 & i & -i \\ -i & -1 & 1 \end{bmatrix},$$

$$B_2 = \begin{bmatrix} 2-i & 2i-1 & i+1 \\ 2i+1 & -i-2 & i-1 \\ -2i+1 & i-2 & -i-1 \\ i+2 & -2i-1 & -i+1 \end{bmatrix}, C_1 = \begin{bmatrix} 8i & -8 & -8i & 8 \\ 4 & 4i & -4 & -4i \\ 2+4i & -4+2i & 2-4i & 4-2i \end{bmatrix},$$

$$C_2 = \begin{bmatrix} 11i & 44i-11 & -44 \\ 22 & 22i+88 & 88i \\ -11i & 44i+11 & 44 \\ -22 & -22i-88 & -88i \end{bmatrix}, A_3 = \begin{bmatrix} 5i+2 & 5-2i & -2+5i & 2i+5 \\ 2i-5 & 5i+2 & -2i-5 & -2+5i \\ 4i & 4 & -4i & -4 \end{bmatrix},$$

$$B_3 = \begin{bmatrix} -i & -i+2 & -1 \\ -2 & -2-4i & 2i \\ -2i & 4-2i & -2 \\ 1 & 1+2i & -i \end{bmatrix}, A_4 = \begin{bmatrix} -2i-3 & -3i+2 & 2i+3 \\ -i & 1 & i \\ -3 & -3i & 3 \end{bmatrix},$$

$$C_3 = \begin{bmatrix} 3i & 3 & -3 & -3i \\ 3 & -3i & 3i & -3 \\ -3 & 3i & -3i & 3 \\ -3i & -3 & 3 & 3i \end{bmatrix},$$

$$B_4 = \begin{bmatrix} 7i & -i & -2 \\ -7 & -3 & 2i \\ -7i & i & 2 \\ 7 & 3 & -2i \end{bmatrix}, C_4 = \begin{bmatrix} 4-2i & -2+4i & 2+2i \\ 2+4i & -4-2i & -2+2i \\ -2-4i & 4+2i & 2-2i \end{bmatrix},$$

$$C_c = \frac{1}{21} \begin{bmatrix} -1130-502i & -1344+612i & -2798-1250i \\ -1808-688 & -1398+834i & -2942-1538i \\ -1154-946i & -1488+624i & -2654-1394i \end{bmatrix}. \tag{36}$$

Let us find a solution to the system (7) with the given above matrices by Algorithm 1.

1. Thanks to Lemma 1.10, we calculate the Moore-Penrose inverses. So,

$$A_1^{\dagger} = \frac{1}{32} \begin{bmatrix} 1-i & -1-i & -2i \\ 1+i & 1-i & 2 \\ -1-i & -1+i & -2 \\ -1+i & 1+i & 2i \end{bmatrix}, B_1^{\dagger} = \frac{1}{44} \begin{bmatrix} -11i & 11i & -11 & 11 \\ 39 & 41 & 20-i & 20+i \\ 7-i & 1+i & 5+3i & 3-3i \end{bmatrix},$$

$$A_2^{\dagger} = \frac{1}{12} \begin{bmatrix} -i & 1 & -1 & i \\ 1 & i & -i & -1 \\ -1 & -i & i & 1 \end{bmatrix}, B_2^{\dagger} = \frac{1}{12} \begin{bmatrix} 1 & -i & i & 1 \\ -i & -1 & -1 & i \\ 1-i & -1-i & -1+i & 1+i \end{bmatrix},$$

$$A_3^{\dagger} = \frac{1}{80} \begin{bmatrix} -2i & -2 & 2-5i \\ 2 & -2i & 5+2i \\ -2i & -2 & 2+5i \\ 2 & -2i & -5+2i \end{bmatrix}, B_3^{\dagger} = \frac{1}{70} \begin{bmatrix} i & -2 & 2i & 1 \\ 2+i & -2+4i & 4+2i & 1-2i \\ -1 & -2i & -2 & i \end{bmatrix},$$

$$A_4^{\dagger}=\frac{1}{69}\begin{bmatrix}-3+2i & i & -3\\ 2+3i & 1 & 3i\\ 3-2i & -i & 1\end{bmatrix},B_4^{\dagger}=\frac{1}{792}\begin{bmatrix}-35i & -21 & 35i & 21\\ 47i & -51 & -47i & 51\\ -52 & -48i & 52 & 48i\end{bmatrix}.$$

Then,

$$A=\frac{1}{2}\begin{bmatrix}2+5i & 5-2i & 1+8i & 12+9i\\ -5+2i & 2+5i & -8+i & -9+12i\\ 4i & 4 & 4-8i & -8+4i\end{bmatrix},$$

$$B=\frac{1}{22}\begin{bmatrix}-52-31i & 10-135i & -31+52i\\ 8+9i & -10+25i & 9-8i\\ -9+8i & -25-10i & 8+9i\\ 31-52i & 135+10i & -52-31i\end{bmatrix},$$

$$C=\frac{1}{3}\begin{bmatrix}-11-3i & 9-7i & 6+4i\\ -1-3i & 3+i & 2i\\ -9+3 & 3-9i & 6\end{bmatrix},D=\begin{bmatrix}0 & -2i & -2\\ 0 & -2 & 2i\\ 0 & 2i & 2\\ 0 & 2 & -2i\end{bmatrix},$$

$$N=\frac{1}{7}\begin{bmatrix}4+4i & -4-2i & -10-4i\\ 4-4i & -2+4i & -4+10i\\ -4-4i & 4+2i & 10+4i\\ -4+4i & 2-4i & 4-10i\end{bmatrix},M=\frac{1}{3}\begin{bmatrix}-4-2i & 4-2i & 2+2i\\ -2+4i & -2-4i & 2-2i\\ 0 & 0 & 0\end{bmatrix},S=0$$

$$E=\frac{1}{84}\begin{bmatrix}19931-108289i & 236509-68427i & -108289-19931i\\ 110417+16211i & 77995+79015i & 16211-110417i\\ 74624+106424i & -138224+255672i & 106424-74624i\end{bmatrix}.$$

2. Confirm that (11) and (14) are true for given matrices.

3. Finally, by (29) and (30), we find that the least-norm of the solution pair X_1 and X_2 to the system (7) is following

$$X_1=\frac{1}{365760}\begin{bmatrix}-11103239+18670545i & -9851419+14002307i & -5154373+3862099i & -4697553+10234559i\\ 26688873+4258681i & 29888893+5510501i & 12048461+4721147i & 17746081+5177967i\\ 6556168+9656066i & 5321848+2196342i & 4452786+10360112i & -6757414+7845632i\\ -17049264-2930378i & -26304464-11113378i & -10244698-3367816i & -7362609-13720296i\end{bmatrix},$$

$$X_2=\frac{1}{1344}\begin{bmatrix}2052-963i & 233-1985i & -2159+3481i & -1465-367i\\ -792+2565i & 1901-205i & 317+445i & -221+317i\\ 171+585i & -146+28i & 868-1714i & 146+2884i\end{bmatrix}.$$

Note that Maple 2021 was used to perform the numerical experiment.

4. Conclusion

We have constructed a novel expression of the general solution to system (7) over $\mathbb{C}$ and used this result to explore the least-norm of the general solution to this system when it is solvable. Some particular cases of our system are also discussed. Our results carry the principal results of [22, 32]. To give an algorithm finding the explicit numerical expression of the least-norm of the general solution, it is used the determinantal representations of the MP-inverse recently obtained by one of the authors. The novelty of the conducted research is obtaining necessary and sufficient conditions to exist a solution, its formal representation of by closed formula in terms of generalized inverses, and the construction of an algorithm to find its explicit expression. A numerical example is also given to interpret the results established in this paper.

Conflict of interest

The authors declare that they have not conflicts of interest.

Data Availability

The data used to support the findings of this study are included within the article titled "Solving and Algorithm for Least-Norm General Solution to Constrained Sylvester Matrix Equation". The prior studies (and datasets) are cited at relevant places within the text.

Classification

2000 AMS subject classifications: 15A09, 15A15, 15A24.

Author details

Abdur Rehman[1] and Ivan I. Kyrchei[2*]

1 University of Engineering and Technology, Lahore, Pakistan

2 Pidstrygach Institute for Applied Problems of Mechanics and Mathematics, NAS of Ukraine, Lviv, Ukraine

*Address all correspondence to: ivankyrchei26@gmail.com

References

[1] Shahzad A, Jones BL, Kerrigan EC, Constantinides GA. An efficient algorithm for the solution of a coupled Sylvester equation appearing in descriptor systems. Automatica. 2011;**47**: 244-248. DOI: 10.1016/j.automatica.2010.10.038

[2] Syrmos VL, Lewis FL. Coupled and constrained Sylvester equations in system design. Circuits, Systems, and Signal Processing. 1994;**13**(6):663-694. DOI: 10.1007/BF02523122

[3] Varga A. Robust pole assignment via Sylvester equation based state feedback parametrization: Computer-aided control system design (CACSD). IEEE International Symposium. 2000;**57**: 13-18. DOI: 10.1109/CACSD.2000.900179

[4] Syrmos VL, Lewis FL. Output feedback eigenstructure assignment using two Sylvester equations. IEEE Transaction on Automatic Control. 1993;**38**:495-499. DOI: 10.1109/9.210155

[5] Li RC. A bound on the solution to a structured Sylvester equation with an application to relative perturbation theory. SIAM Journal on Matrix Analysis and Application. 1999;**21**(2): 440-445

[6] Darouach M. Solution to Sylvester equation associated to linear descriptor systems. Systems and Control Letters. 2006;**55**:835-838. DOI: 10.1016/j.sysconle.2006.04.004

[7] Zhang YN, Jiang DC, Wang J. A recurrent neural network for solving Sylvester equation with time-varying coefficients. IEEE Transcation on Neural Networks. 2002;**13**(5):1053-1063. DOI: 10.1109/TNN.2002.1031938

[8] Terán FD, Dopico FM. The solution of the equation $XA + AX^T = 0$ and its application to the theory of orbits. Linear Algebra and its Application. 2011; **434**:44-67. DOI: 10.1016/j.laa.2010.08.005

[9] Dehghan M, Hajarian M. An efficient iterative method for solving the second-order Sylvester matrix equation $EVF^2 - AVF - CV = BW$. IET Control Theory and Applications. 2009;**3**: 1401-1408. DOI: 10.1049/iet-cta.2008.0450

[10] Ding F, Chen T. Gradient based iterative algorithms for solving a class of matrix equations. IEEE Transaction on Automatic Control. 2005;**50**(8): 1216-1221. DOI: 10.1109/TAC.2005.852558

[11] Dmytryshyn A, Futorny V, Klymchuk T, Sergeichuk VV. Generalization of Roth's solvability criteria to systems of matrix equations. Linear Algebra and its Application. 2017; **527**:294-302. DOI: 10.1016/j.laa.2017.04.011

[12] He ZH, Wang QWA. Real quaternion matrix equation with applications. Linear and Multilinear Algebra. 2013;**61**(6):725-740. DOI: 10.1080/03081087.2012.703192

[13] He ZH, Wang QW. The η-bihermitian solution to a system of real quaternion matrix equation. Linear and Multilinear Algebra. 2014;**62**(11): 1509-1528. DOI: 10.1080/03081087.2013.839667

[14] Kyrchei II. Explicit representation formulas for the minimum norm least squares solutions of some quaternion

matrix equations. Linear Algebra and its Applications. 2013;**438**(1):136-152. DOI: 10.1016/j.laa.2012.07.049

[15] Kyrchei II. Explicit determinantal representation formulas for the solution of the two-sided restricted quaternionic matrix equation. Journal of Applied Mathematics and Computing. 2018;**58** (1–2):335-365. DOI: 10.1007/s12190-017-1148-6

[16] Rehman A, Wang QW. A system of matrix equations with five variables. Applied Mathematics and Computation. 2015;**271**:805-819. DOI: 10.1016/j.amc.2015.09.066

[17] Rehman A, Wang QW, Ali I, Akram M, Ahmad MO. A constraint system of generalized Sylvester quaternion matrix equations. Adv. Appl. Clifford Algebr. 2017;**27**(4):3183-3196. DOI: 10.1007/s00006-017-0803-1

[18] Rehman A, Wang QW, He ZH. Solution to a system of real quaternion matrix equations encompassing η-Hermicity. Applied Mathematics and Computation. 2015;**265**:945-957. DOI: 10.1016/j.amc.2015.05.104

[19] Rehman A, Akram M. Optimization of a nonlinear hermitian matrix expression with application. Univerzitet u Nišu. 2017;**31**(9):2805-2819. DOI: 10.2298/FIL1709805R

[20] Wang QW, Qin F, Lin CY. The common solution to matrix equations over a regular ring with applications. Indian Journal of Pure and Applied Mathematics. 2005;**36**(12):655-672

[21] Wang QW, Rehman A, He ZH, Zhang Y. Constraint generalized Sylvester matrix equations. Automatica. 2016;**69**:60-64. DOI: 10.1016/j.automatica.2016.02.024

[22] Bao Y. Least-norm and extremal ranks of the Least Square solution to the quaternion matrix equation $AXB = C$ subject to two equations. Algebra Colloq. 2014;**21**(3):449-460. DOI: 10.1142/S100538671400039X

[23] Wang QW, Chang HX, Lin CY. P-(skew)symmetric common solutions to a pair of quaternion matrix equations. Applied Mathematics and Computation. 2008;**195**:721-732. DOI: 10.1016/j.amc.2007.05.021

[24] Li H, Gao Z, Zhao D. Least squares solutions of the matrix equation $AXB + CYD = E$ with the least norm for symmetric arrowhead matrices. Applied Mathematics and Computation. 2014;**226**:719-724. DOI: 10.1016/j.amc.2013.10.065

[25] Wang QW, van der Woude JW, Chang HX. A system of real quaternion matrix equations with applications. Linear Algebra and its Application. 2009;**431**(12):2291-2303. DOI: 10.1016/j.laa.2009.02.010

[26] Peng YG, Wang X. A finite iterative algorithm for solving the least-norm generalized (P, Q) reflexive solution of the matrix equations $A_iXB_i = C_i$. Journal of Computational Analysis and Applications. 2014;**17**(3):547-561

[27] Yuan S, Liao A. Least squares Hermitian solution of the complex matrix equation $AXB + CXD = E$ with the least norm. Journal of Franklin Institute. 2014;**351**(11):4978-4997. DOI: 10.1016/j.jfranklin.2014.08.003

[28] Trench WF. Minimization problems for (R, S)-symmetric and (R, S)-skew symmetric matrices. Linear Algebra and its Applications. 2004;**389**:23-31. DOI: 10.1016/j.laa.2004.03.035

[29] Trench WF. Characterization and properties of matrices with generalized symmetry or skew symmetry. Linear Algebra and its Applications. 2004;**377**: 207-218. DOI: 10.1016/j.laa.2003.07.013

[30] Trench WF. Characterization and properties of (R,S)-symmetric, (R,S)-skew symmetric and (R,S)-conjugate matrices. SIAM Journal on Matrix Analysis and Application. 2005;**26**: 748-757. DOI: 10.1137/S089547980343134X

[31] Marsaglia G, Styan GPH. Equalities and inequalities for ranks of matrices. Linear and Multilinear Algebra. 1974;**2**: 269-292. DOI: 10.1080/03081087408817070

[32] Wang QW, Li CK. Ranks and the least-norm of the general solution to a system of quaternion matrix equations. Linear Algebra and its Application. 2009; **430**:1626-1640. DOI: 10.1016/j.laa.2008.05.031

[33] Wang QW, Chang HX, Ning Q. The common solution to six quaternion matrix equations with applications. Applied Mathematics and Computation. 2008;**198**:209-226. DOI: 10.1016/j.amc.2007.08.091

[34] Tian Y. Solvability of two linear matrix equations. Linear and Multilinear Algebra. 2000;**48**:123-147. DOI: 10.1080/03081080008818664

[35] Wang QW, Wu ZC, Lin CY. Extremal ranks of a quaternion matrix expression subject to consistent systems of quaternion matrix equations with applications. Applied Mathematics and Computation. 2006;**182**:1755-1764. DOI: 10.1016/j.amc.2006.06.012

[36] Wang QW. A system of matrix equations and a linear matrix equation over arbitrary regular rings with identity. Linear Algebra and its Application. 2004;**384**:43-54. DOI: 10.1016/j.laa.2003.12.039

[37] Wensheng C. Solvability of a quaternion matrix equation. Applied Mathematics, Journal of Chinese Universities, Serie B. 2002;**17**(4): 490-498. DOI: 10.1007/s11766-996-0015-2

[38] Artidiello S, Cordero A, Torregrosa JR, Vassileva MP. Generalized inverses estimations by means of iterative methods with memory. Mathematics. 2019;**8**:2. DOI: 10.3390/math8010002

[39] Guo W, Huang T. Method of elementary transformation to compute Moore–Penrose inverse. Applied Mathematics and Computation. 2010; **216**:1614-1617. DOI: 10.1016/j.amc.2010.03.016

[40] Sayevand K, Pourdarvish A, Machado JAT, Erfanifar R. On the calculation of the Moore-Penrose and Drazin inverses: Application to fractional calculus. Mathematics. 2021;**9**:2501. DOI: 10.3390/math9192501

[41] Bapat RB, Bhaskara KPS, Prasad KM. Generalized inverses over integral domains. Linear Algebra and its Applications. 1990;**140**:181-196. DOI: 10.1016/0024-3795(90)90229-6

[42] Stanimirovic PS. General determinantal representation of pseudoinverses of matrices. Matematichki Vesnik. 1996;**48**:1-9

[43] Kyrchei II. Analogs of the adjoint matrix for generalized inverses and corresponding Cramer rules. Linear and Multilinear Algebra. 2008;**56**(4): 453-469. DOI: 10.1080/03081080701352856

[44] Kyrchei II. Determinantal representation of the Moore-Penrose inverse matrix over the quaternion skew field. Journal of Mathematical Sciences. 2012;**108**(1):23-33. DOI: 10.1007/s10958-011-0626-x

[45] Kyrchei II. Determinantal representations of the Drazin and W-weighted Drazin inverses over the quaternion skew field with applications. In: Griffin S, editor. Quaternions: Theory and Applications. New York: Nova Sci Publ; 2017. pp. 201-275

[46] Kyrchei II. Determinantal representations of the quaternion weighted Moore-Penrose inverse and its applications. In: Baswell AR, editor. Advances in Mathematics Research 23. New York: Nova Sci Publ; 2017. pp. 35 96

[47] Kyrchei II. Cramer's rule for generalized inverse solutions. In: Kyrchei II, editor. Advances in Linear Algebra Research. New York: Nova Sci Publ; 2015. pp. 79-132

[48] Kyrchei II. Analogs of Cramer's rule for the minimum norm least squares solutions of some matrix equations. Applied Mathematics and Computation. 2012;**218**(11):6375-6384. DOI: 10.1016/j.amc.2011.12.004

[49] Kyrchei II. Determinantal representations of solutions and hermitian solutions to some system of two-sided quaternion matrix equations. Journal of Mathematics. 2018; **2018**:6294672. DOI: 10.1155/2018/6294672

[50] Kyrchei II. Cramer's rules of η-(skew-)Hermitian solutions to the quaternion Sylvester-type matrix equations. Adv. Appl. Clifford Algebr. 2019;**29**(3):56. DOI: 10.1007/s00006-019-0972-1

[51] Kyrchei II. Determinantal representations of solutions to systems of two-sided quaternion matrix equations. Linear and Multilinear Algebra. 2021;**69**(4):648-672. DOI: 10.1080/03081087.2019.1614517

[52] Rehman A, Kyrchei II, Ali I, Akram M, Shakoor A. The general solution of quaternion matrix equation having η-skew-Hermicity and its Cramer's rule. Mathematical Problems in Engineering. 2018;**2018**:7939238. DOI: 10.1155/2019/7939238

[53] Rehman A, Kyrchei II, Ali I, Akram M, Shakoor A. Explicit formulas and determinantal representation for η-skew-Hermitian solution to a system of quaternion matrix equations. Univerzitet u Nišu. 2020;**34**(8):2601-2627. DOI: 10.2298/FIL2008601R

[54] Rehman A, Kyrchei II, Ali I, Akram M, Shakoor A. Constraint solution of a classical system of quaternion matrix equations and its Cramer's rule. Iranian Journal of Science and Technology, Transactions A: Science. 2021;**45**(3):1015-1024. DOI: 10.1007/s40995-021-01083-7

[55] Tian Y. Equalities and inequalities for traces of quaternionic matrices. Algebras Groups Geometry. 2002;**19**(2): 181-193